STUDENT STUDY GUIDE & SELECTED SOLUTIONS MANUAL

DAVID D. REID
Eastern Michigan Universty

PHYSICS
for Scientists and Engineers
THIRD EDITION

FISHBANE | GASIOROWICZ | THORNTON

PEARSON

Prentice
Hall

Upper Saddle River, NJ 07458

Associate Editor: Christian Botting
Senior Editor: Erik Fahlgren
Editor-in-Chief, Science: John Challice
Vice President of Production & Manufacturing: David W. Riccardi
Executive Managing Editor: Kathleen Schiaparelli
Assistant Managing Editor: Becca Richter
Production Editor: Elizabeth Klug
Supplement Cover Manager: Paul Gourhan
Supplement Cover Designer: Joanne Alexandris
Manufacturing Buyer: Ilene Kahn

© 2005 Pearson Education, Inc.
Pearson Prentice Hall
Pearson Education, Inc.
Upper Saddle River, NJ 07458

The author and publisher of this book have used their best efforts in preparing this book. These efforts include the development, research, and testing of the theories and programs to determine their effectiveness. The author and publisher make no warranty of any kind, expressed or implied, with regard to these programs or the documentation contained in this book. The author and publisher shall not be liable in any event for incidental or consequential damages in connection with, or arising out of, the furnishing, performance, or use of these programs.

Printed in the United States of America

10 9 8 7 6 5 4 3 2 1

ISBN 0-13-100070-5

Pearson Education Ltd., *London*
Pearson Education Australia Pty. Ltd., *Sydney*
Pearson Education Singapore, Pte. Ltd.
Pearson Education North Asia Ltd., *Hong Kong*
Pearson Education Canada, Inc., *Toronto*
Pearson Educación de Mexico, S.A. de C.V.
Pearson Education—Japan, *Tokyo*
Pearson Education Malaysia, Pte. Ltd.

TABLE OF CONTENTS

PREFACE

This study guide is designed to assist you in your study of the fascinating and challenging world of physics using volume 1 of the second edition of *Physics for Scientists and Engineers*, by Fishbane, Gasiorowicz, and Thornton. To do this I have provided a Chapter Review, which consists of a comprehensive, but brief, review of every section in the text. Numerous solved examples and exercises appear throughout each chapter review. The Examples in this study guide follow the excellent pedagogical format of the Examples in the text. Together with the Chapter Review, each chapter contains a list of objectives, a practice quiz, a glossary of key terms and phrases, a table of important formulas, and a table that reviews the units of the new quantities introduced.

In addition to the above materials that I have provided, you will also find Practice Problems by Carl Adler (East Carolina University) and Selected Solutions to End of Chapter Problems as modified from the *Instructor's Solutions Manual*. Taken together, the information in this study guide, when used in conjunction with the main text, should enhance your ability to master the many concepts and skills needed to understand physics, and therefore, the world around you. Work hard, and most importantly, have fun doing it!

I am indebted to many for helping me to complete this work. Most directly, I thank Mr. Christian Botting of Prentice Hall for selecting me to do this project. I also wish to acknowledge Dr. Michael Ottinger of Missouri Western State College. He provided an excellent review of the physics content and made countless valuable suggestions.

David D. Reid
Eastern Michigan University
May, 2004

To my family
Thanks for all of your support!

CHAPTER 1

TOOLING UP

Chapter Objectives

After studying this chapter, you should

1. be able to use scientific notation.
2. be able to convert from one unit to another.
3. be able to perform calculations while keeping proper account of the number of significant figures.
4. be able to check equations by performing a dimensional analysis.
5. be able to represent vectors both graphically and mathematically.
6. know the difference between scalars and vectors.
7. be able to determine the magnitude and direction of a vector.
8. be able to determine the components of a vector.
9. be able to write a vector in unit vector notation.
10. be able to add and subtract vectors both graphically and algebraically.

Chapter Review

This chapter prepares you to study physics by covering basic concepts and mathematical techniques that are vital to your ability to understand and apply physics. It is particularly important that you thoroughly learn how to work with systems of units (section 1–2) and vectors (section 1–6).

1–1   A Little Background

Knowledge in physics progresses through an interplay between experimental measurements and theoretical formulations. This interplay occurs on many different scales because different formulations of physics apply within different regimes. Despite the need for these different formulations, there are only a few basic laws of physics. These basic laws, and how to apply them, are at the core of almost every branch of engineering.

1–2   The Fundamental Physical Quantities and Their Units

A very useful way of writing numerical values is to use **scientific notation**. In this notation, a value is written as a number of order unity (meaning that only one digit is left of the decimal point) times the appropriate power of 10, the *order of magnitude*. Scientific notation provides a consistent way to represent numbers of widely varying sizes, from a very small number such as 9.1×10^{-31}, to a very large number such as 6.0×10^{24}. Furthermore, using this notation often simplifies numerical calculations. Other benefits to using scientific notation will be explored later.

Example 1–1 Write the following lengths using scientific notation.

(a) 0.0025 in **(b)** 12,100 in **(c)** 451 in **(d)** 8.00 in **(e)** 0.0000359 in

Answer: (a) 2.5×10^{-3} in **(b)** 1.21×10^4 in **(c)** 4.51×10^2 in **(d)** 8.00×10^0 in **(e)** 3.59×10^{-5} in

When the value is already of order unity, as with part (d), the power of 10 is usually dropped.

As a starting point in the use of physical quantities to describe the world we adopt *length*, *time*, and *mass* as three fundamental quantities that are used to define other derived quantities. We devise a system of units for these quantities so that we can specify how much length, time, or mass we have. The system of units used in this book is called **SI**, which stands for Système International. In this system, the unit of length is the *meter* (m), the unit of time is the *second* (s), and the unit of mass is the *kilogram* (kg). This system of units is still sometimes referred to by its former name, the mks system. Many of the units in SI are represented using the convenient system of metric prefixes (see Table 1–4 in the textbook).

Some other systems of units are still in common use. The *cgs system* is based on the centimeter (cm), gram (g), and second (s). The *British engineering system* is based on the inch (in), pound (lb), and second (s). Be aware that the pound is not a unit of mass; it is a unit of force. You will study the concept of force in Chapter 4. Because different systems of units are common, It is important to know how to convert from one unit to another.

A conversion can be accomplished using a **conversion factor** that is constructed from a primary conversion equation (how much of a quantity in one unit equals that same quantity in another unit). A conversion factor is a ratio of equal quantities written such that, when multiplied by a quantity, the undesired unit algebraically cancels leaving only the desired unit. This concept is best illustrated by example.

Example 1–2 A typical cardboard box provided by moving companies measures 1.50 ft × 1.50 ft × 1.33 ft. Determine the volume (V) of clothes that you can pack into this box in cubic meters.

Setting it up: Each dimension of the box is given explicitly in the problem. Respectively, we can take these to be the values of the depth (d), width (w), and height (h) of the box.

Strategy: The volume of the box is given by $V = dwh$. Let's first calculate the volume in the given units, and then convert the volume to cubic meters.

Working it out: The volume of the box is

$$V = 1.50 \text{ ft} \times 1.50 \text{ ft} \times 1.33 \text{ ft} = 2.9925 \text{ ft}^3$$

Primary conversion equations are given in the inside front cover of the textbook. There, we see that 1 ft = 0.3048 m. Because we need to convert from feet to meters, our conversion factor should be written such that 1 ft is in the denominator. Thus,

$$V = 2.9925 \text{ ft}^3 \left(\frac{0.3048 \text{ m}}{1 \text{ ft}} \right)^3 = 0.0847 \text{ m}^3$$

Notice that the entire conversion factor was cubed because there are three factors of length.

What do you think?   To what power would you raise the conversion factor if you were converting the bottom area of the box from the given units to SI?

Practice Quiz

1. Which of the following numbers is the proper scientific notation for 25,300?

   (a) 2.53×10^4 **(b)** 25.3×10^3 **(c)** 2.53×10^3 **(d)** 0.253×10^5

2. The number 7.4×10^5 is equivalent to which of the following?

   (a) 7.4   **(b)** 740   **(c)** 7,400   **(d)** 740,000

1–3   Accuracy and Significant Figures

With every measurement of a physical quantity, there is an associated **uncertainty**. The uncertainty tells us how accurate the measured value is. For example, the length of a pencil might be given as $18.3 \pm 0.1 \, \text{cm}$, where 0.1 cm is the uncertainty in the measured value (or central value) of 18.3 cm. It is also common for uncertainties to be given as a percentage of the measured value. For the pencil, we have that $0.1/18.3 = 0.005$, so the percentage uncertainty is 0.5%. Therefore, we can also state the result of the measurement as $18.3 \, \text{cm} \pm 0.5\%$.

Because there is uncertainty in every measured quantity, when quoting values we must be careful to accurately state the precision to which we know the value. We keep track of the precision through the number of digits used to state the value. Digits that are reliably known are called **significant figures**. The basic facts concerning significant figures are

* *Zeros*: Zeros that are written only to locate the decimal point (set the order of magnitude) are not significant figures. All other zeros are counted as significant figures. Some zeros can be ambiguous, this will be discussed in the Tips section.
* *Multiplication and Division*: The number of significant figures in the result of a multiplication or division equals the number of significant figures in the factor containing the fewest significant figures.
* *Addition and Subtraction*: The significant figures in the result of an addition or subtraction are located only in the *places* (hundreds, ones, tenths, etc.) that are reliably known for *every* value in the sum.

Example 1–3 A calculation involves the addition of two measured distances $d_1 = 1250$ m, and $d_2 = 336$ m. If each measurement is given to three significant figures, what is the result of the calculation?

Setting it up: We are told that each measurement is given to three significant figures. Therefore, the last zero in d_1 is not significant, it only locates the decimal point.

Strategy: Because not every digit listed is significant in both values, we must be careful to follow the rule for addition and subtraction by looking at the places occupied by those digits that are significant.

Working it out: If we blindly add the two numbers we get a value of 1586. However, because the one's place of 1250 m (the 0) is not significant, the one's place of the result cannot be significant, even though the 6 is significant in 336 m. Therefore, we must round our answer off to the ten's place. So,

$$d_1 + d_2 = 1250 \text{ m} + 336 \text{ m} = 1590 \text{ m}$$

What do you think? You may wonder about the fact that there is no significant figure in the thousand's place of 336 m, why then, is it okay to consider the thousand's place of the result to be significant?

Practice Quiz

3. Assuming that every nonzero digit is significant, consider the following product of numbers: $1.34 \times 10.75 \times 0.042$. Which answer is correct to the proper number of significant figures?

    (a) 0.6   **(b)** 0.61   **(c)** 0.605   **(d)** 0.60501

4. Assuming that only nonzero digits are significant, consider the following sum of numbers: $1700 + 338 + 13$. Which answer is correct to the proper number of significant figures?

    (a) 2051   **(b)** 2050   **(c)** 2100   **(d)** 2000

To avoid excessive round-off error, you should round to the proper number of significant figures only at the very end of a calculation. In Example 1–3, if the distance calculated is only an intermediate step in a longer calculation, then the value 1586 m should be used in the subsequent steps. In general, keep at least one additional digit for values calculated in intermediate steps (in Example 1–2, I needed two additional digits in the volume calculation). An even better approach is not to calculate intermediate values numerically, but to carry through the *formulas* inserting numerical values only at the end.

1–4 — 1–5   Dimensional Analysis and Estimates

The **dimension** of a quantity tells us what *type* of quantity it is. When indicating the dimension of a quantity only, we use capital letters enclosed in brackets. Thus, the dimension of length is represented by $[L]$, time by $[T]$, and mass by $[M]$. Equations relating physical quantities must be dimensionally consistent. It is useful to perform a dimensional analysis on any equation about which you are unsure. If the equation is not dimensionally consistent, it cannot be a correct equation. The rules are simple:

      * Two quantities can be added or subtracted only if they are of the same dimension.

      * Two quantities can be equal only if they are of the same dimension.

    Notice that only the dimension needs to be the same, not the units. It is perfectly valid to write 12 in = 1 ft because both quantities are lengths, $[L] = [L]$. However, it is not valid to write x inches = t seconds because the quantities have different dimensions: $[L] \neq [T]$.

Example 1–4 Given that the quantities x (m), v (m/s), a (m/s^2), and t (s) are measured in the units shown in parentheses, perform a dimensional analysis on **(a)** $x = 2vt$ and **(b)** $v = at + t/x$.

Strategy: We need to write out the dimension of each quantity in the equation and check whether the rules concerning dimensional analysis are obeyed.

Working it out:

(a) The equation written in terms of dimensions is $[L] = [LT^{-1}] \times [T]$. Because the right-hand side is algebraically equivalent to the left-hand side, we have $[L] = [L]$. So, the equation is dimensionally correct.

(b) The equation written in terms of dimensions is $[LT^{-1}] = [LT^{-2}] \times [T] + [L^{-1}T]$. The right-hand side reduces to $[LT^{-2}] \times [T] + [L^{-1}T] = [LT^{-1}] + [L^{-1}T]$. These two terms are not of equal dimension and therefore, cannot be added. So, the equation for part (b) is not a valid equation.

What do you think? For part (a), what effect did the 2 have on the analysis? What does your answer imply about the ability of dimensional analysis to confirm that an equation is correct?

Practice Quiz

5. If speed v has units of m/s, distance d has units of m, and time t has units of s, which of the following expressions is dimensionally correct?

    (a) $v = t/d$ **(b)** $t = vd$ **(c)** $d = v/t$ **(d)** $t = d/v$

We have just seen that dimensional analysis provides a quick check to determine if an equation might be correct. You can conduct a quick check on the numerical value of a quantity by performing an order-of-magnitude calculation. To do this type of estimate, make a reasonable guess for the values of the quantities that are needed to determine the result you seek. Here, reasonable means within the correct order-of-magnitude. The result of the calculation, when performed using these guesses, should provide a reasonable estimate of the more accurate value. Typically, such estimates are rounded to just one significant figure.

Example 1–5 Suppose you need to perform the conversion of Example 1–2, but you don't have the conversion factor available. Perform an order-of-magnitude estimate of the conversion of the volume to cubic meters.

Setting it up/Strategy: We know the volume in cubic feet is 2.9925 ft^3. So, we need a reasonable estimate of the conversion factor.

Working it out: One meter is comparable to one yard. Most of us have learned that 3 feet equals 1 yard, and we can estimate that 3 ft ≈ 1 m . So,

$$V = 2.9925 \ \text{ft}^3 \left(\frac{1 \ \text{m}}{3 \ \text{ft}} \right)^3 = 0.1 \ \text{m}^3$$

to 1 significant figure. Thus, we'd expect the correct answer to be close to this value. If a more careful calculation gives a value of, say, 10 m³ (two orders of magnitude larger), this would make us suspicious of a possible error in the calculation. The correct value to 1 significant figure of 0.08 m³, however, is right in the ballpark of our estimate.

What do you think?   Construct a different argument to estimate the conversion factor. How does your estimate compare to the correct value?

1–6   Scalars and Vectors

For many quantities used in physics, a simple number, together with its units, suffices to specify the quantity. Such quantities are called **scalar** quantities. Other quantities, however, require a directional specification in addition to a numerical value. Such quantities are called **vector** quantities. The ability to work with vector quantities is extremely important in physics. Study this topic thoroughly.

The numerical value of a vector quantity is called its **magnitude** (always a positive number). Together, the magnitude and the direction completely specify the vector. Vector quantities will be distinguished from scalar quantities by use of an arrow, such as $\vec{V}$. The magnitude of the vector $\vec{V}$ is a scalar quantity; there are two common notations for the magnitude of a vector, V and $|\vec{V}|$, with V being more common.

Vector quantities can be represented both geometrically and algebraically. Geometrically, a vector is an arrow in a coordinate system. A two-dimensional vector is shown below.

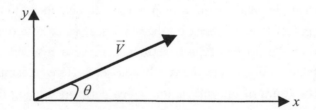

In the geometric representation, the magnitude of the vector is represented by the length of the arrow. Its direction is specified by its orientation with respect to the axes of the coordinate system; it is customary, but not necessary, to use the angle made with the *x*-axis. Algebraically, one way to specify a vector is by stating the numerical values of its magnitude (with units) and direction (either in words or using angles). Another way to specify a vector is by using components and unit vectors.

A **unit vector** is a dimensionless vector whose magnitude equals 1. Generally, unit vectors are used to indicate specific directions. A unit vector will be distinguished from other vectors by having a ^ over it. When a unit vector is multiplied by a scalar, the result is a vector whose magnitude equals the absolute value of the scalar and whose direction is the same as that of the unit vector if the scalar is positive, or the opposite of the unit vector if the scalar is negative. Therefore, a vector $\vec{V}$ of magnitude V can also be written as $V\hat{V}$, where $\hat{V}$ is a unit vector in the direction of $\vec{V}$. Our most common application of unit vectors will be for specifying the *x*- and *y*-directions. The unit vector for the *x* direction is $\hat{i}$, and the unit vector for the *y* direction is $\hat{j}$.

Working with vector quantities can often be simplified by resolving them into **components**. In two dimensions, a vector has two components, one corresponding to its extent along the *x*-axis and the other corresponding to its extent along the *y*-axis. In the figure below, the vector $\vec{V}$ is resolved into two components, V_x and V_y.

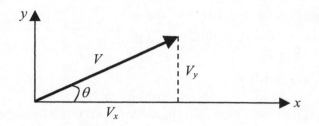

Notice that the magnitude of the vector, V, and the two components form a right triangle; hence, we can use the trigonometric functions to relate all the relevant quantities

$$V_x = V\cos\theta, \quad V_y = V\sin\theta, \quad \theta = \tan^{-1}\left(\frac{V_y}{V_x}\right)$$

As you can see from these relations, the components can be positive, negative, or zero.

Be careful to note that the third of the preceding equations, for the angle, provides only a *reference angle* for the vector with respect to one of its axes. To know precisely what the direction is, you must also account for the signs of the components or, equivalently, the quadrant of the coordinate system in which the vector lies. (This last point will be illustrated in the solved example.) If you know the components and want the magnitude, then you can use the Pythagorean theorem

$$V = \sqrt{V_x^{\,2} + V_y^{\,2}}$$

As mentioned previously, the combination of components and unit vectors provides another way to algebraically represent a vector quantity. For example, if an arbitrary vector $\vec{A}$ has an *x*-component of 5, $A_x = 5$, and a *y*-component of –8, $A_y = -8$, then, we can write this vector in terms of unit vectors by multiplying its components by the corresponding unit vectors and summing them:

$$\vec{A} = A_x\hat{i} + A_y\hat{j} = 5\hat{i} - 8\hat{j}$$

The quantities $A_x\hat{i}$ and $A_y\hat{j}$ are sometimes called the *component vectors* of $\vec{A}$.

Example 1–6 You're trying to find State Park, but you're lost. You ask someone for directions and she tells you that you can get there by first traveling 250 m west then 310 m north. If you wish to use a vector $\vec{R}$ to specify this location, **(a)** what are the components of this vector, **(b)** what is its magnitude, **(c)** what is its direction, and **(d)** how would you write it in unit vector notation?

Setting it up:

It is helpful to draw a diagram. The diagram shows a coordinate system with the various directions labeled. The location is indicated by the open circle, and the arrow is the vector we wish to use to specify this location. The angle θ is used to specify the direction. It is also helpful to list, and assign variables to, the quantities that are given in the problem, as well as to those we need to find.

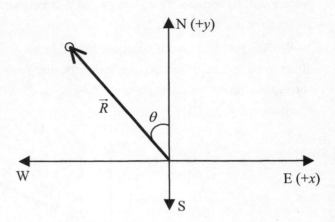

Given: $d_W = 250$ m, $d_N = 310$ m

Find: (a) R_x, R_y, **(b)** R, **(c)** θ, **(d)** $\vec{R}$

Strategy: Because our coordinate axes are oriented to be north-south and east-west, we can tell that the given distances are closely related to the components we seek for part (a). Once we determine that relationship and get the components, the remaining calculations, for parts (b) – (d) follow from the vector formulas written previously.

Working it out:

(a) West corresponds to the negative x-direction; therefore we can make the identification $R_x = -d_W$. Similarly, because north corresponds to the positive y-direction, we can state $R_y = d_N$. Thus,

$$R_x = -250 \text{ m}, \quad R_y = 310 \text{ m}$$

(b) Knowing both the x and y components of the vector, we can obtain the magnitude using the Pythagorean theorem

$$R = \sqrt{R_x^2 + R_y^2} = \sqrt{(-250 \text{ m})^2 + (310 \text{ m})^2} = 398 \text{ m}$$

(c) To determine the angle θ shown in the diagram, we note that the right triangle formed by the vector and its components allows us to use trigonometric functions to calculate this angle.

$$\theta = \tan^{-1}\left(\frac{|R_x|}{R_y}\right) = \tan^{-1}\left(\frac{250 \text{ m}}{310 \text{ m}}\right) = 38.9°$$

(d) Knowing both the x and y components of the vector, we can write down this vector using unit vectors

$$\vec{R} = -(250 \text{ m})\hat{i} + (310 \text{ m})\hat{j}$$

What do you think? The above solution assumed that the given information was accurate to three significant figures. What would the answers be if we only assume two significant figures?

Practice Quiz

6. Taking north to be the $+y$ direction and east to be $+x$, calculate the x and y components (x, y) of a vector whose magnitude is 15 m and is directed $40°$ south of west.

    (a) (11 m, 9.6 m)   **(b)** (-9.6 m, 11m)   **(c)** (-13 m, -11 m)   **(d)** (9.6 m, 13 m)   **(e)** (-11 m , -9.6 m)

7. A certain vector has a y component of 17 in arbitrary units. If its direction is 153° counterclockwise from the +x-direction, what is its magnitude?

    (a) 44   **(b)** 37   **(c)** 16   **(d)** -44   **(e)** 40

Mathematical Operations with Vectors

For now, there are two mathematical operations with vectors that you must learn, *scalar multiplication* and *vector addition*. If an arbitrary vector $\vec{A}$ is multiplied by a scalar c, the result is a vector of magnitude $|c|A$ pointing in the direction $\hat{A}$ if c is positive, or $-\hat{A}$ if c is negative. Also, each component of $\vec{A}$ is multiplied by c. To summarize, if $\vec{B} = c\vec{A}$, then

$$B = |c|A, \quad \hat{B} = \text{sgn}(c)\hat{A}, \quad \vec{B} = (cA_x)\hat{i} + (cA_y)\hat{j}$$

where $\text{sgn}(c)$ denotes the algebraic sign of the scalar c.

There are two approaches to vector addition and subtraction; a graphical method based on geometry, and a component method based on algebra. For precise calculations, you will predominantly use the component method, but the graphical approach can be invaluable for picturing the physical situation being described.

To picture the graphical addition of two vectors $\vec{A}$ and $\vec{B}$ to form a third vector $\vec{C}$, that is, $\vec{C} = \vec{A} + \vec{B}$, imagine traveling along the two vectors. First you travel along $\vec{A}$ from its tail to its head, and then you immediately travel along $\vec{B}$ from its tail to its head. Your net trip, from where you started to where you stopped, will be $\vec{C}$.

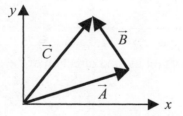

We can summarize this procedure by the following rule:

> *To add two vectors, place the tail of the second vector at the head of the first. The sum of the two then is the vector extending from the tail of the first vector to the head of the second.*

The act of moving vector $\vec{B}$ and placing it at the head of $\vec{A}$ is allowed because vectors are characterized only by their length (magnitude) and orientation (direction), not by their location in the coordinate system. The order in which two vectors are added does not matter, that is, $\vec{A} + \vec{B} = \vec{B} + \vec{A}$.

We can approach vector subtraction in precisely the same way as addition by recalling that subtraction is the addition of the negative of a quantity; that is, the expression $\vec{D} = \vec{A} - \vec{B}$ is equivalent to the expression $\vec{D} = \vec{A} + \left(-\vec{B}\right)$. Therefore, once we form the vector $-\vec{B}$ we can proceed with vector addition as already described. The negative of a vector is formed by rotating the vector 180°, so that you end up with a vector of equal length pointing in the opposite direction as shown below.

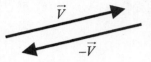

Algebraic vector addition, using components, is carried out by applying the following rule:

> *To add vectors using components, the components of the resultant vector are given by the sum of the corresponding components of the vectors being added.*

To state it mathematically, if $\vec{D} = \vec{A} + \vec{B} + \vec{C}$, then

$$D_x = A_x + B_x + C_x \quad \text{and} \quad D_y = A_y + B_y + C_y$$

Once the components are known, everything else about the vector can be determined, as previously shown in Example 1–6.

For vector subtraction, we once again use the fact that subtraction is the addition of the negative of a vector. Algebraically, the negative of a vector is obtained by changing the sign of each component. Therefore, if vector $\vec{A}$ has components A_x and A_y, then vector $-\vec{A}$ will have components $-A_x$ and $-A_y$. Thus, for the vector difference $\vec{C} = \vec{A} - \vec{B}$,

$$C_x = A_x - B_x \quad \text{and} \quad C_y = A_y - B_y$$

Example 1–7 A vector $\vec{r}$ has components $r_x = -12.0$ and $r_y = 15.0$. Another vector $\vec{s}$ has components $s_x = 9.00$ and $s_y = -4.00$. Determine the magnitude and direction of the sum $\vec{r} + \vec{s}$.

Setting it up:
It is helpful to first sketch the graphical vector addition. The tail of $\vec{s}$ has been placed at the tip of $\vec{r}$. The dashed vector is the sum $\vec{r} + \vec{s}$, which has been called $\vec{d}$. The angle ϕ is used for the direction of $\vec{d}$.

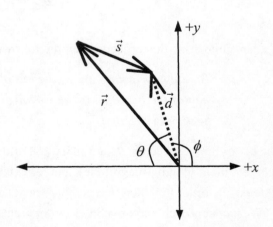

Given: $r_x = -12.0$, $r_y = 15.0$, $s_x = 9.00$, $s_y = -4.00$
Find: d and ϕ

Strategy: We know how to get the magnitude and direction of $\vec{d}$ from its components, so we'll first find the components by using vector addition.

Working it out:

From our sketch of the problem, we expect a negative x component and a positive y component. The x component of $\vec{d}$ is

$$d_x = r_x + s_x = -12.0 + 9.00 = -3.00$$

and the y component of $\vec{d}$ is

$$d_y = r_y + s_y = 15.0 + (-4.00) = 11.0$$

Both results match our expectation. The magnitude of $\vec{d}$ is

$$d = \left(d_x^{\,2} + d_y^{\,2}\right)^{1/2} = \left[(-3.00)^2 + (11.0)^2\right]^{1/2} = 11.4$$

Knowing that the vector lies in the second quadrant, it is easiest to first find the reference angle θ

$$\theta = \tan^{-1}\left(\frac{d_y}{|d_x|}\right) = \tan^{-1}\left(\frac{11.0}{3.00}\right) = 74.7°$$

However, when using an angle to specify the direction of a two-dimensional vector, it is customary to use the angle the vector makes with the positive x axis. Thus,

$$\phi = 180° - \theta = 180° - 74.7° = 105°$$

What do you think? By looking at the components of $\vec{s}$, in what quadrant does it lie? Is your answer consistent with the way $\vec{s}$ is drawn in the sketch?

Practice Quiz

8. A vector is given by $\vec{A} = 3.0\hat{i} - 5.0\hat{j}$. What is the magnitude of the vector $5\vec{A}$?

(a) 15   (b) 29   (c) 10   (d) 40   (e) 25

9. Which of the following sketches correctly represents the vector subtraction $\vec{R_3} = \vec{R_1} - \vec{R_2}$?

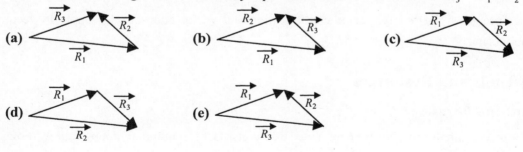

10. If a vector $\vec{Q}$ has x and y components of -3.0 and 5.5 respectively, and vector $\vec{R}$ has x and y components of 9.2 and 4.4 respectively, which vector equals the sum $\vec{Q} + \vec{R}$?

(a) $14.7\hat{i} + 1.4\hat{j}$ (b) $6.2\hat{i} + 9.9\hat{j}$ (c) $3.7\hat{i} + 7.4\hat{j}$ (d) $2.5\hat{i} - 4.8\hat{j}$ (e) $-3.0\hat{i} + 4.4\hat{j}$

11. Suppose a vector $\vec{A}$ has x and y components of 41 and 28, respectively and vector $\vec{B}$ has x and y components of 12 and -18 respectively. Determine the magnitude of the vector $\vec{C}$ if $\vec{C} = \vec{B} + \vec{A}$.

(a) 50   (b) 22   (c) 54   (d) 28   (e) 72

12. Given the vectors $\vec{A}$, $\vec{B}$, and $\vec{C}$ in question 11, what is the direction of $\vec{C}$, relative to the standard +x-direction?

(a) 11°   **(b)** 34°   **(c)** –22°   **(d)** 56°   **(e)** 91°

The quantity that serves as a prototype for all vector quantities is the **displacement vector**. The displacement of an object is the vector from some starting position to a subsequent position. The displacement vector $\vec{s}$ can be written as the difference between two **position vectors**, one for the starting position $\vec{r_1}$ and one for the subsequent position $\vec{r_2}$

$$\vec{s} = \vec{r_2} - \vec{r_1}$$

where a position vector represents the displacement from the origin of a particular coordinate system.

In general, space is three-dimensional; therefore, we will also need to represent vectors in three-dimensional space. Cartesian coordinates in three dimensions adds a third axis, the z-axis, to the familiar x-y coordinate system. The z-axis is perpendicular to the x-y plane such that the positive z-direction is consistent with the following *right-hand rule*:

> *Point the fingers of your right hand along the positive x-axis such that they can curl toward the positive y-axis. Your extended thumb will point along the positive z-axis.*

A three-dimensional vector $\vec{V}$ has three components, V_x, V_y, and V_z such that it can be written

$$\vec{V} = V_x\hat{i} + V_y\hat{j} + V_z\hat{k}$$

where $\hat{k}$ is the unit vector for the z-direction. The magnitude of a three-dimensional vector is given by

$$V = \sqrt{V_x^2 + V_y^2 + V_z^2}$$

All of the rules previously stated for scalar multiplication and vector addition also apply to vectors in three-dimensional space.

Reference Tools and Resources

I. Key Terms and Phrases

scientific notation   a method of writing numbers that consists of a number of order unity times the appropriate power of 10

SI   the internationally adopted standard system of units (based on meters, kilograms, and seconds) for quantitatively measuring quantities

conversion factor   a factor (equal to 1) that multiplies a quantity to convert its value to another unit

uncertainty   the amount by which the measured value of a quantity may vary

significant figures   the digits in the numerical value of a quantity that are accurately known

dimension of a quantity   the fundamental type of a quantity such as length, mass, or time

scalar   a numerical value with appropriate units

vector a mathematical quantity having both magnitude and direction (with appropriate units)

magnitude of a vector the full numerical value of the quantity being represented

unit vector a dimensionless vector of unit magnitude

component of a vector the part of a vector associated with a specific direction

displacement vector the vector from one position to another

position vector the displacement vector from the origin of a coordinate system

II. Important Equations

Name/Topic	Equation	Explanation
Components of a vector $\vec{V}$	$V_x = V\cos\theta,\ V_y = V\sin\theta$	The components of a vector from its magnitude and direction. The angle θ is measured from the $+x$ direction.
Reference angle	$\theta = \tan^{-1}\left(\dfrac{V_y}{V_x}\right)$	The reference angle gives the direction of a vector with respect to the coordinate axes.
Magnitude of a vector	$V = \sqrt{V_x^{\,2} + V_y^{\,2}}$	The magnitude of a vector from its x and y components.
Unit vectors	$\vec{V} = V_x\hat{i} + V_y\hat{j}$	A two-dimensional vector written in terms of unit vectors.
Scalar multiplication	$c\vec{A} = (cA_x)\hat{i} + (cA_y)\hat{j}$	Scalars multiply each component of a vector.
Vector addition	$\vec{A} \pm \vec{B} = (A_x \pm B_x)\hat{i} + (A_y \pm B_y)\hat{j}$	Vector addition and subtraction with components.

III. Know Your Units

Quantity	Dimension	SI Unit
Displacement	$[L]$	m
Unit vector	—	dimensionless

IV. Tips
Significant Figures
In section 1–3 it was briefly mentioned that a zero can be ambiguous as to whether it is a significant figure. If I tell you that I walked 250 m, does that last zero simply locate the decimal point, or am I really saying that the distance I walked is closer to 250 m than it is to 249 m or 251 m? An additional benefit of using scientific notation is that it removes any such ambiguities because the numerical prefactor is only written with the significant figures. For example, if the number 3700 has only one significant figure, we write it as 4×10^3; if it has two, we write 3.7×10^3; if it has three, we write 3.70×10^3; and if it has four significant figures we write 3.700×10^3.

Dimensional Analysis
You should be aware that, typically, arguments of mathematical functions are dimensionless. Angles, for example, are dimensionless, as can be seen by the equation for the length of a circular arc, $s = r\theta$, where θ is in radians; hence, angular measures such as radians and degrees signify only how we choose to measure the angle. The trigonometric functions, therefore, such as sine, cosine, and tangent, are applied to dimensionless quantities. Other examples of dimensionless functions are $\log(x)$, $\ln(x)$, and their inverse functions 10^x and $\exp(x)$.

Vector Addition
The component method of vector addition described in this chapter is very powerful and always works. However, the method can sometimes be cumbersome, and increasing the number of calculations provides greater opportunities for some sort of mistake to slip in. Consider Example 1–7, I gave the components of the vectors that needed to be added, but what if I had given their magnitudes and (angular) directions instead? That would have required their components to be calculated, adding at least four more calculations (maybe more depending on which angles were given) before the actual vector addition could begin. For such cases, the law of cosines can often provide a more direct route to the result than the component method. The law of cosines is illustrated below.

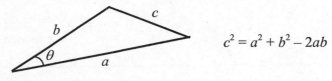

$$c^2 = a^2 + b^2 - 2ab$$

For the vectors in Example 1–7, if we were given the magnitudes and directions of $\vec{r}$ and $\vec{s}$, we could immediately draw the following triangle:

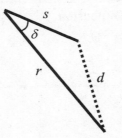

The law of cosines would then yield the magnitude d with just one additional step

$$d = \left[r^2 + s^2 - 2rs \cos(\delta) \right]^{1/2}$$

Practice Problems

1. What is the decimal equivalent of 3.14×10^3?

2. What is the product of 2.8×10^{-4} and 3.14×10^3?

3. $10.2 \times 7.2 =$

4. $27.1 / 5.03 =$

5. $2.712 + 10.5 =$

6. What is the volume, in cubic cm, of a sphere with a radius of 2 cm?

7. A vector has x component 7.1 and y component –7.9. To the nearest tenth of a unit, what is its magnitude?

8. To the nearest tenth of a degree, what is the direction of the vector in Problem 7? (Report your results in the format $-180°$ to $180°$.)

9. What is the volume in cubic cm of a cylinder with a diameter of 1.6 cm and a height of 2.7 cm?

10. 113 miles (exactly) is how many meters (exactly)?

Selected Solutions to End of Chapter Problems

5. Taking the cube root of a number is equivalent to raising it to the one-third power. So,

$$\sqrt[3]{10^{21}} = \left(10^{21}\right)^{1/3} = 10^{21/3} = \boxed{10^7}$$

If we now square this number we get

$$\left(10^7\right)^2 = 10^{7 \cdot 2} = \boxed{10^{14}}$$

13. Average density is mass divided by volume. Taking the Moon to be spherical, the volume of a sphere is $V = 4\pi R^3/3$, where R is the radius. Therefore, the density, ρ, is given by $\rho = M/V$.

$$\rho = \frac{M}{V} = \frac{3M}{4\pi R^3} = \frac{3\left(7.35 \times 10^{22} \text{ kg}\right)}{4\pi \left(1.74 \times 10^3 \text{ km}\right)^3} = 3.331 \times 10^{12} \text{ kg/km}^3$$

Converting this to the desired units, and rounding to the correct number of significant digits, gives

$$\rho = 3.331 \times 10^{12} \frac{kg}{km^3} \left(\frac{10^3 g}{kg}\right)\left(\frac{km}{10^5 cm}\right)^3 = \boxed{3.33 \text{ g/cm}^3}$$

19. The volume of the box is the product of each dimension. The percent uncertainty of a product is approximately equal to the sum of the percent uncertainties of each factor. So, for each dimension we first find the percent uncertainty:

$w = 0.75 \pm 0.02$ m $= 0.75$ m $\pm (0.02/0.75)(100\%) = 0.75$ m $\pm 3\%$;

$l = 0.5 \pm 0.1$ m $= 0.5$ m $\pm (0.1/0.5)(100\%) = 0.5$ m $\pm 20\%$;

$h = 0.582 \pm 0.058$ m $= 0.582$ m $\pm (0.058/0.582)(100\%) = 0.582$ m $\pm 10\%$.

Then, the volume is given by

$$V = wlh = (0.75 \text{ m})(0.5 \text{ m})(0.582 \text{ m}) \pm (3\% + 20\% + 10\%).$$

Rounding both the value and the percent uncertainty to the appropriate number of significant digits, we get

$$V = \boxed{0.2 \text{ m}^3 \pm 30\%} = 0.2 \pm 0.06 \text{ m}^3 \approx 0.2 \pm 0.1 \text{ m}^3.$$

Note that we need to round off the error from 0.06 m^3 to 0.1 m^3 in order to match V, since the last digit in the main figure has to be aligned with the first digit in its error.

31. (a) Written as a dimensional equation we have that $[Kq^2/hc] = [K] [q^2] / [h] [c]$. Because the quantity K is dimensionless, we can drop it from the equation to get $[q^2/hc] = [q^2] / [h][c]$. In order for the entire quantity to be dimensionless, it must be that $[q^2] = [q]^2 = [h][c]$. From problem 30 we know that $[h] = [ML^2 T^{-1}]$ and $[c] = [LT^{-1}]$; thus, $[q]^2 = [ML^2 T^{-1} \times LT^{-1}] = [ML^3 T^{-2}]$. So, finally, we have

$$[q] = \boxed{[M^{1/2} L^{3/2} T^{-1}]}$$

(b) We know $[q^2]$ from part (a), so here we need to divide that result by a length. This gives,

$$[q^2/R] = [ML^3 T^{-2}] / [L] = \boxed{[ML^2 T^{-2}]}$$

53. If we take $\hat{i}$ as east and $\hat{j}$ as north, the four displacements, in paces, are

$$4\hat{j}, \quad 6\cos(45°)\hat{i} + 6\sin(45°)\hat{j}, \quad 2\hat{i}, \quad -5\hat{i}$$

The resultant displacement is their sum:

$$\vec{R} = 4\hat{j} + 6\cos(45°)\hat{i} + 6\sin(45°)\hat{j} + 2\hat{i} + -5\hat{i} = \boxed{1.2\hat{i} + 8.2\hat{j} \text{ paces}}$$

As a magnitude and a direction, we have

$$R = \sqrt{(1.2 \text{ paces})^2 + (8.2 \text{ paces})^2} = 8.3 \text{ paces}$$

and

$$\theta = \tan^{-1}\left(\frac{8.2}{1.2}\right) = 82°$$

Therefore, the displacement can also be stated as $\boxed{8.3 \text{ paces at } 82° \text{ north of east}}$.

61. (a) Reading from the diagram given in the problem, we determine that

$$\vec{A} = -4\hat{i} + 2\hat{j}, \quad \vec{B} = -\hat{i} + 4\hat{j}, \quad \vec{C} = 2\hat{i} + 2\hat{j}, \quad \vec{D} = 5\hat{i} - 3\hat{j}$$

(b) <u>Algebraically</u>

$$2\vec{A} + \vec{C} - \vec{D} = 2\left(-4\hat{i} + 2\hat{j}\right) + \left(2\hat{i} + 2\hat{j}\right) - \left(5\hat{i} - 3\hat{j}\right) = (-8 + 2 - 5)\hat{i} + (4 + 2 + 3)\hat{j} = \boxed{-11\hat{i} + 9\hat{j}}$$

$$\vec{B} + \vec{C}/2 = \left(-\hat{i} + 4\hat{j}\right) + \left(2\hat{i} + 2\hat{j}\right)/2 = (-1 + 1)\hat{i} + (4 + 1)\hat{j} = \boxed{5\hat{j}}$$

$$\vec{D} - \vec{B} = \left(5\hat{i} - 3\hat{j}\right) - \left(-\hat{i} + 4\hat{j}\right) = (5 + 1)\hat{i} + (-3 - 4)\hat{j} = 6\hat{i} - 7\hat{j}$$

$$\therefore \quad \left|\vec{D} - \vec{B}\right| = \sqrt{6^2 + (-7)^2} = \boxed{9.2}$$

<u>Graphically</u>

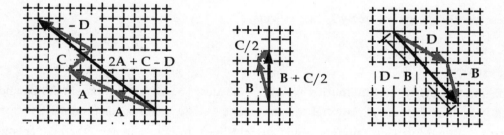

Answers to Practice Quiz

1. (a) **2.** (d) **3.** (b) **4.** (c) **5.** (d) **6.** (e) **7.** (b) **8.** (b) **9.** (e) **10.** (b) **11.** (c) **12.** (a)

Answers to Practice Problems

1. 3140   **2.**   0.88
3. 73   **4.**   5.39
5. 13.2   **6.**   30 cm³
7. 10.6   **8.**   −48.1°
9. 5.4 cm³   **10.** 181855.872

CHAPTER 2

STRAIGHT-LINE MOTION

Chapter Objectives

After studying this chapter, you should

1. know the difference between distance and displacement.
2. know the difference between speed and velocity.
3. know the difference between velocity and acceleration.
4. be able to calculate displacements, velocities, and accelerations using the equations of one-dimensional motion.
5. be able to interpret x-versus-t and v-versus-t plots for motion with both constant velocity and constant acceleration.
6. be able to describe the motion of freely falling objects.

Chapter Review

This chapter begins the study of **kinematics** which is the description of motion. In kinematics, we describe motion using three primary quantities, displacement, velocity, and acceleration. Here, we focus on motion along just one dimension (along a single straight line). In later chapters, the concepts learned here will be applied to motion in higher dimensions.

2–1 Displacement

In kinematics, we often seek to describe how an object moves as a function of time. If we let $\vec{x}_1$ be the position of a particle at time t_1, and $\vec{x}_2$ be the position at time t_2, then the displacement of the particle (its change in position) is given by

$$\Delta \vec{x} = \vec{x}_2 - \vec{x}_1$$

where the symbol Δ means "the change in..." Even though position and displacement are vector quantities, in one-dimensional motion it is sufficient to work with their components, Δx for the displacement, and x for the position. For a one-dimensional vector, the absolute value of the component equals the magnitude of the vector, and the algebraic sign of the component tells you its direction. Notice that the displacement of an object does not depend on the path taken; it only depends on the two relevant positions.

Example 2–1 From your apartment, you leave your favorite parking spot and drive 4.83 km east on Main Street to go to the grocery store. After shopping, you go back home by traveling west on Main Street and find that your favorite parking spot is still available. **(a)** What distance do you travel during this trip? **(b)** What is your displacement?

Setting it up: It is helpful to draw a diagram. We adopt a coordinate system with east being to the right and west being to the left. The origin is set at the parking spot; and s represents the distance between the parking spot and the store.

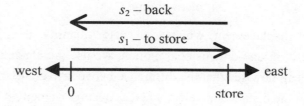

From the statement of the problem, and our coordinate system, we can write the following information.
Given: $s_1 = s_2 = 4.83$ km, $x_2 = x_1 = 0$;   **Find**: (a) s_{tot}, (b) Δx

Strategy: For this problem, we must note the distinction between distance and displacement. For the distance, we must consider the entire length of the path taken. For the displacement, we ignore the path and focus on the initial and final positions only.

Working it out:

(a) The path taken has two length segments: s_1, going to the store, and s_2, coming back home. The total length of travel, s, is the sum of these two segments.
$$s = s_1 + s_2 = 4.83 \text{ km} + 4.83 \text{ km} = 9.66 \text{ km}$$

(b) Here, we notice that the initial and final positions are the same. Subtract the initial position from the final position to get the displacement.
$$\Delta x = x_2 - x_1 = 0 \text{ m} - 0 \text{ m} = 0 \text{ m}$$

What do you think? Would the final answers to parts (a) and (b) be different if we place the origin at the store rather than the parking spot?

Practice Quiz

1. If you walk exactly four times around a quarter-mile track, what is your displacement?

   (a) one mile   **(b)** half a mile   **(c)** one-quarter mile   **(d)** zero

2–2 — 2–3 Speed, Velocity, and Acceleration

Speed and Velocity

The concepts of distance and displacement relate to the fact that an object moves from one position to another. Additionally, an important part of describing motion is to specify how rapidly an object moves. One way to do this is to quote an **average speed**,

$$\text{average speed} = \frac{\text{total distance}}{\text{elapsed time}}$$

Thus, average speed is a positive scalar quantity.

Another, sometimes more appropriate, way to describe the rate of motion is to quote a **velocity**. The velocity of an object is its rate of change in position. The average velocity of an object is given by

$$v_{av} = \frac{\text{displacement}}{\text{elapsed time}} = \frac{\Delta x}{\Delta t}$$

Because velocity depends on displacement, which is a vector quantity, then velocity is also a vector quantity. As with displacement, in one-dimensional motion, we use the algebraic sign of the velocity to tell us its direction. Thus, we can now see that the difference between average speed and average velocity is that average speed relates to the distance traveled, whereas average velocity relates to the displacement. Therefore, average speed tells us about the average rate of motion over the entire path taken and contains no directional information. The SI unit of speed (and velocity) is m/s.

Example 2–2 For the trip described in Example 2–1, if it took $\Delta t_1 = 10.0$ min to drive to the store and $\Delta t_2 = 12.0$ min to drive back home, calculate the average speed and average velocity for the trip.

Setting it up: Having the results of Example 2-1, we can write out the following information:
Given: $s = 9.66$ km, $\Delta x = 0$, $\Delta t_1 = 10.0$ min, $\Delta t_2 = 12.0$ min ;   **Find:** ave speed, v_{av}

Strategy: We can directly apply the definitions of average speed and average velocity. However, we have a mix of units in this problem, so we should remember to convert everything to SI.

Working it out: Let's determine the elapsed time in seconds.

$$\Delta t = \Delta t_1 + \Delta t_2 = 10.0 \text{ min} + 12.0 \text{ min} = 22.0 \text{ min} = 22.0 \text{ min}\left(\frac{60 \text{ s}}{\text{min}}\right) = 1320 \text{ s}$$

The distance traveled is $s = 9.66$ km $= 9.66 \times 10^3$ m. Therefore, the average speed is

$$\text{ave speed} = \frac{s}{\Delta t} = \frac{9.66 \times 10^3 \text{m}}{1320 \text{ s}} = 7.32 \text{ m/s}$$

For the average velocity, we have

$$v_{av} = \frac{\Delta x}{\Delta t} = \frac{0 \text{ m}}{1320 \text{ s}} = 0 \text{ m/s}$$

What do you think? If you considered the trips to the store and back home separately, would the average speeds and average velocities be different? (Be careful about signs.)

Some situations require more than just the average rate of motion. Often, we require the velocity that an object has at a specific instant in time; this velocity is called the **instantaneous velocity**. The instantaneous velocity, v, can be defined in terms of the average velocity measured over an infinitesimally small elapsed time,

$$v = \lim_{\Delta t \to 0} \frac{\Delta x}{\Delta t} = \frac{dx}{dt}$$

Thus, the instantaneous velocity would be the velocity that an object has right at $t = 2.0$ s, for example, instead of the average velocity over a time period $\Delta t = 2.0$ s. The magnitude of the instantaneous velocity of an object (how fast it is going) is its **instantaneous speed** $|v|$.

In general, average and instantaneous velocities will have very different values; however, during constant-velocity motion, the average velocity over any time interval equals the instantaneous velocity at any time $v_{av} = v_0$. Thus, the definition of average velocity also serves as an equation that describes constant-velocity motion. For this special case, the equation is often written as

$$\Delta x = v_0 \Delta t \quad \text{or} \quad x(t) = v_0 t + x_0$$

where the subscript "0" refers to the initial value of the quantity.

Practice Quiz

2. In a coordinate system in which east is the positive direction and west is the negative direction, you take a total time of 105 seconds to walk 20 m west, then 10 m east, followed by 15 m west. With what average speed have you walked?

   (a) 0.24 m/s   **(b)** 0.43 m/s   **(c)** –0.24 m/s   **(d)** –0.43 m/s   **(e)** 0 m/s

3. For the information given in question 3, with what average velocity have you walked?

   (a) 0.24 m/s   **(b)** 0.43 m/s   **(c)** –0.24 m/s   **(d)** –0.43 m/s   **(e)** 0 m/s

4. How long does it take a person on a bicycle to travel exactly 1 km if she rides at a constant velocity of 20 m/s?

   (a) 20,000 s   **(b)** 0.020 s   **(c)** 50 s   **(d)** 20 s   **(e)** 5.0 s

Graphing the Motion

The motion of an object is often analyzed graphically. In order to do this analysis, we must know how to obtain information from these graphs. For now, we focus on graphs of position as a function of time. In general, plots of position-versus-time will be curved. Regardless of the shape of the curve, however, information about the velocity of the motion can be determined from the graph. Any two points on the graph can be connected by a straight line. The slope of this connecting line equals the average velocity of the object over the corresponding time interval. Also, for any given point on the curve there is a line called the *tangent line* that intersects the curve at that point. The slope of the tangent line at a point equals the instantaneous velocity of the object at the corresponding time.

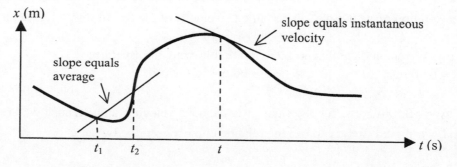

For the special case of constant-velocity motion, the equation $x(t) = v_0 t + x_0$ shows that we expect the x-versus-t graph to be linear. The slope of the line will give us the velocity of the motion and the x-intercept gives the initial position.

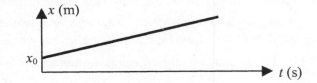

Acceleration

Another very important concept is the **acceleration** of an object. Acceleration is the rate at which an object's velocity is changing. It doesn't matter if the object is slowing down, speeding up, or changing direction; if there is any change in the speed and/or direction of motion of an object, it is accelerating. Sometimes we need only the average rate of change of velocity, or *average acceleration*,

$$a_{av} = \frac{v_2 - v_1}{t_2 - t_1} = \frac{\Delta v}{\Delta t}$$

At other times we may need the *instantaneous acceleration*

$$a = \lim_{\Delta t \to 0} \frac{\Delta v}{\Delta t} = \frac{dv}{dt}$$

As an object accelerates from an initial velocity v_1 to a final velocity v_2, it may have many different values of instantaneous acceleration along the way. Basically, the average acceleration tells us what constant acceleration would produce the same velocity change Δv in the same amount of elapsed time Δt. Whereas, the instantaneous acceleration tells us the value only at a particular instant in time. The SI unit of acceleration is m/s^2.

The change in velocity that results from a particular acceleration depends on how the velocity and acceleration relate to each other. In general, when the velocity and acceleration have opposite signs, the object will slow down. When the velocity and acceleration have the same sign (whether both are negative or positive), the object will speed up. The relationship between acceleration and position follows from the definition of instantaneous acceleration

$$a = \frac{d}{dt}(v) = \frac{d}{dt}\left(\frac{dx}{dt}\right) = \frac{d^2x}{dt^2}$$

Example 2–3 A particle has an instantaneous velocity of 6.00 m/s. Over the next 4.00 s it experiences an average acceleration of -3.00 m/s^2. Determine its velocity at the end of this 4-s time interval.

Setting it up: We can reduce this problem to the following information:
Given: $v_0 = 6.00$ m/s, $\Delta t = 4.00$ s, $a_{av} = -3.00$ m/s^2;   **Find:** v

Strategy: We need to solve for the final instantaneous velocity; and because we are given the average acceleration, we should first try to use the relationship between the two.

Working it out: Using the variable names assigned above, the expression for average acceleration is

$$a_{av} = \frac{v - v_0}{\Delta t}$$

This equation can be rearranged to give

$$v = v_0 + a_{av}\Delta t = 6.00 \text{ m/s} + \left(-3.00 \text{ m/s}^2\right)\left(4.00 \text{ s}\right) = -6.00 \text{ m/s}$$

What do you think? It is a simple mathematical solution, but do you understand the motion it describes? What is the significance of the fact that the initial velocity and the average acceleration have different signs? Notice, however, that the final velocity has the same sign as the average acceleration (is opposite to that of the initial velocity); what must have happened?

Practice Quiz

5. The velocity of a particle as a function of time is given by $v(t) = \left(5.0 \text{ m/s}^3\right)t^2 - 2.0 \text{ m/s}$. What is the acceleration of this particle (in m/s^2) at time $t = 2.2$ s ?

  (a) –2.0   **(b)** 22   **(c)** 5.0   **(d)** 11   **(e)** 20

2–4 Motion with Constant Acceleration

An important special case of accelerated motion occurs when the acceleration is constant. Conceptually, this means that the rate at which the velocity changes is the same at every instant in time during the motion. As was the case with velocity, when the acceleration of an object is constant, the average acceleration equals the instantaneous acceleration. This means that the definition of average acceleration provides an equation that can be used to describe motion with constant acceleration, $a = \Delta v / \Delta t$. In this context, it is customary to simply call the final time t ($t_2 = t$) and to define the initial time to be $t_1 = 0$; hence, $\Delta t = t_2 - t_1 = t$. With this definition, the other initial quantities, x_1 and v_1, are the values that correspond to $t = 0$ and are labeled accordingly: $v_1 \rightarrow v_0$ and $x_1 \rightarrow x_0$. With these modifications, we can rewrite the average velocity equation as

$$v = at + v_0$$

Notice, however, that the preceding equation does not involve position. This suggests that to completely describe motion with constant acceleration we will need more than just this one equation. In fact, we use four equations, relating the different quantities of interest, to describe this type of motion. The other three equations are

$$v_{av} = \tfrac{1}{2}\left(v_0 + v\right); \quad x = \tfrac{1}{2}at^2 + v_0 t + x_0; \quad \text{and} \quad v^2 = v_0^2 + 2a\left(x - x_0\right)$$

These four equations contain all the information needed to describe motion with constant acceleration.

Example 2–4 An object moves with a constant acceleration of 1.75 m/s^2 and reaches a velocity of 7.80 m/s after 3.20 s. How far does it travel during those 3.20 s?

Setting it up: We can reduce this problem to the following information:

Given: $a = 1.75$ m/s^2, $v = 7.80$ m/s, $t = 3.20$ s ;   Find: $x - x_0$

Strategy: Two of the four equations for constant acceleration contain the displacement $x - x_0$. Looking at those equations, we need the initial velocity v_0 in order to solve for the displacement. It would be easy to assume that $v_0 = 0$, but the problem doesn't say that. So, let us first try to solve for v_0, and then calculate the displacement.

Working it out:
Each of the four equations contain v_0, so we must compare the given information with each of those expressions. Doing this comparison shows that the first of the four equations listed above gives

$$v_0 = v - at$$

which can be solved because we are given v, a, and t. However, let's not solve it numerically. Leaving v_0 in equation form is a better way to handle round-off error, as mentioned in Chapter 1.

Going back to the two equations that involve the displacement, we see that by substituting in the expression for v_0, we can use either of these equations to solve for $x - x_0$. Let us choose $x = \frac{1}{2}at^2 + v_0 t + x_0$, and write it as

$$x - x_0 = \frac{1}{2}at^2 + [v - at]t$$

which reduces to

$$x - x_0 = -\frac{1}{2}at^2 + vt$$

Using this final expression gives

$$x - x_0 = -\frac{1}{2}(1.75 \text{ m/s}^2)(3.20 \text{ s})^2 + (7.80 \text{ m/s})(3.20 \text{ s}) = 16.0 \text{ m}$$

The final expression we obtained is completely general within the context of motion with constant acceleration. If you wish, you can add this equation as a fifth equation to be used for describing this type of motion.

What do you think? We never did solve for v_0. What is its value? Is it in the same direction, or opposite direction, as the acceleration?

Practice Quiz

6. The driver of a car that is initially moving at 25.0 m/s west applies the brakes until he is going 15.0 m/s west. If the car travels 13.5 m while slowing at a constant rate, what is its acceleration and in what direction is it?

    (a) 0.675 m/s^2, east   **(b)** 20.0 m/s^2, west   **(c)** 14.8 m/s^2, east   **(d)** 22.2 m/s^2, west

7. If a car cruises at 11.3 m/s for 75.0 s, then uniformly speeds up until, after 45.0 s, it reaches a speed of 18.5 m/s, what is the car's displacement during this motion?

    (a) 671 m   **(b)** 1.52 km   **(c)** 3.58 km   **(d)** 848 m   **(e)** 0.160 km

2–5 Freely Falling Objects

It is an experimental fact that when air resistance is negligible, objects near Earth's surface fall with a constant acceleration $\vec{g}$. The symbol g represents the magnitude of this acceleration, which has an approximate value of

$$g = 9.80 \text{ m/s}^2$$

The direction of this acceleration is downward. This direction is commonly taken to be the negative direction along the axis defining the coordinate system. In this case, the acceleration, a, is given by $a = -g$, but notice that the value of g is always positive. Objects undergoing this type of motion, when gravity is the only important influence, are said to be in **free fall**. In some cases, it may be more convenient to choose downward as the positive direction, in which case $a = +g$.

Because the acceleration is constant, the four equations for motion with constant acceleration, from the previous section, also apply to free fall motion. The only differences in how the equations are written are that (a) we will use the variable y for vertical motion (instead of x), and (b) we will take downward to be the negative direction substituting $-g$ for the acceleration. The four equations now become

$$v = -gt + v_0; \quad v_{av} = \tfrac{1}{2}(v_0 + v); \quad y = -\tfrac{1}{2}gt^2 + v_0 t + y_0; \quad \text{and} \quad v^2 = v_0^2 - 2g(y - y_0)$$

There are a few useful facts that can be deduced from the above four equations.

* When an object launched vertically upward reaches the top of its path (its maximum height), its instantaneous velocity is zero, even though its acceleration continues to be 9.80 m/s^2 downward.

* An object launched upward from a given height takes an equal amount of time to reach the top of its path as it takes to fall from the top of its path back to the height from which it was launched.

* The velocity an object has at a given height, on its way up, is equal and opposite to the velocity it will have at that same height on its way back down.

Example 2–5 An entertainer is learning to juggle balls thrown very high. One of the balls is thrown vertically upward from 1.80 m above the floor with an initial velocity of 4.92 m/s. If he fails to catch the ball and it hits the floor, how long is it in the air?

Setting it up:

It is helpful to draw a diagram. The diagram shows the coordinate system with down as negative and the origin at the initial height of the ball. The ball is launched vertically upward and falls straight back down. Remember that the upward and downward paths of the ball are actually along the same line. The problem reduces to the following information:

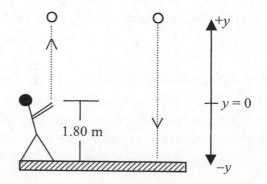

Given: $y_0 = 0$ m, $v_0 = 4.92$ m/s, $y = -1.80$ m

Find: t

Strategy: We are asked to find the time when the ball is at a certain position. So, we need an expression that relates time to position, initial velocity, and acceleration (which we know to be $-g$ because it is free fall). By examining the four equations for free-fall motion, we notice that $y = -\frac{1}{2}gt^2 + v_0 t + y_0$ satisfies these requirements.

Working it out: Now that we know which equation to use, we need to solve for the time t. The equation is quadratic in t, so we put it in standard form and use the fact that $y_0 = 0$

$$\left(-\frac{g}{2}\right)t^2 + v_0 t - y = 0$$

We can now apply the quadratic formula to get the solutions

$$t = \frac{-v_0}{-g} \pm \frac{\sqrt{v_0^2 - 2gy}}{-g}$$

$$t = \frac{-4.92 \text{ m/s}}{-9.80 \text{ m/s}^2} \pm \frac{\sqrt{(4.92 \text{ m/s})^2 - 2(9.80 \text{ m/s}^2)(-1.80 \text{ m})}}{-9.80 \text{ m/s}^2} = 0.5020 \text{ s} \mp 0.7870 \text{ s}$$

Between these two solutions to the equation, it is clear that the correct answer for this problem must be the one corresponding to the positive time. Therefore,

$$t = 0.5020 \text{ s} + 0.7870 \text{ s} = 1.29 \text{ s}$$

What do you think? Why should the negative solution be rejected? To what does it correspond?

Practice Quiz

8. A ball is thrown vertically upward from a height of 2.00 m above the ground with a speed of 17.3 m/s. If the ball is caught by the same person at the same height, how long is the ball in the air?
    (a) 1.76 s   **(b)** 4.00 s   **(c)** 3.53 s   **(d)** 8.65 s

9. Can an object that is moving upward be in free fall?
    (a) No, because an object cannot be falling if it is moving upward
    (b) Yes, because Earth's gravity sometimes pushes objects upward
    (c) No, because Earth's gravity never pushes objects upward
    (d) Yes, as long as gravity is the only force acting on it

*2–6 Integration and Motion in One Dimension

We have seen previously that displacement, velocity, and acceleration are related to each other by differentiation. This fact implies that they are also related by **integration**, because differentiation and integration are inverse processes. So, if we let $g(t) = df(t)/dt$, then

$$f(t) = \int g(t)dt$$

Geometrically, integration corresponds to calculating the area under a curve.

Therefore, displacement is the integral of velocity over time

$$v = \frac{dx}{dt} \quad \Rightarrow \quad \Delta x = \int_{t_1}^{t_2} v(t)dt$$

or, equivalently, the area under the velocity-versus-time curve. Similarly, velocity is the integral of acceleration over time

$$a = \frac{dv}{dt} \quad \Rightarrow \quad v_2 - v_1 = \int_{t_1}^{t_2} a(t)dt$$

or, equivalently, the area under the acceleration-versus-time curve. One of the benefits of using integration techniques is that with them, you can study motion that varies arbitrarily with time as opposed to the special cases of constant velocity and constant acceleration that we have studied in this chapter. A number of useful integrals are listed in Appendix IV of the textbook.

Reference Tools and Resources

I. Key Terms and Phrases

kinematics the branch of physics that describes motion

average speed distance divided by elapsed time

velocity the rate of change of position with time

instantaneous velocity the velocity at a specific instant in time

instantaneous speed equals the magnitude of the instantaneous velocity

acceleration the rate of change of velocity with time

free fall the motion of an object subject only to the influence of gravity

integration the inverse mathematical process of differentiation

II. Important Equations

Name/Topic	Equation	Explanation
Displacement	$\Delta x = x_2 - x_1$	Displacement is the change in position of an object.
Average velocity	$v_{av} = \frac{\Delta x}{\Delta t}$	Average velocity is the displacement divided by the elapsed time.
Instantaneous velocity	$v = \frac{dx}{dt}$	Instantaneous velocity is the rate of change in position.

Constant velocity motion	$x(t) = v_0 t + x_0$	When velocity is constant, position changes linearly with time.
Average acceleration	$a_{av} = \dfrac{\Delta v}{\Delta t}$	Average acceleration is the velocity change divided by the elapsed time.
Instantaneous acceleration	$a = \dfrac{dv}{dt} = \dfrac{d^2 x}{dt^2}$	Instantaneous acceleration is the first derivative of v and the second derivative of x with respect to t.
Motion with constant acceleration	$v = at + v_0$	Velocity changes linearly with time.
	$v_{av} = \tfrac{1}{2}(v_0 + v)$	Average velocity.
	$x = \tfrac{1}{2} at^2 + v_0 t + x_0$	Position in terms of a and t.
	$v^2 = v_0^2 + 2a(x - x_0)$	Velocity squared in terms of x–x_0.
Integration and motion	$\Delta x = \displaystyle\int_{t_1}^{t_2} v(t)\,dt$	Displacement is the integral of velocity over time
	$v_2 - v_1 = \displaystyle\int_{t_1}^{t_2} a(t)\,dt$	Velocity is the integral of acceleration over time.

III. Know Your Units

Quantity	Dimension	SI Unit
Velocity (v)	$[LT^{-1}]$	m/s
Acceleration (a)	$[LT^{-2}]$	m/s^2

IV. Tips

The Equations for Constant Acceleration

The equations for constant acceleration, including the equation derived in Example 2–4, are often written as a set of equations in which each equation has a key physical quantity missing. When written this way, the equation for average velocity is substituted into the definition of average velocity $v_{av} = \Delta x / \Delta t$ to produce an equation for the displacement. Thus, the set of five equations becomes

$$v = at + v_0; \quad x - x_0 = \tfrac{1}{2}(v_0 + v)t; \quad x - x_0 = \tfrac{1}{2}at^2 + v_0 t; \quad v^2 = v_0^2 + 2a(x - x_0); \quad x - x_0 = -\tfrac{1}{2}at^2 + vt$$

Taking them in the order listed, the first equation is missing displacement $(x - x_0)$, the second is missing acceleration a, the third is missing final velocity v, the fourth is missing time t, and the fifth is missing initial velocity v_0. From this point of view, if an important quantity is unknown, you have an equation that does not require it.

Graphing Motion with Constant Acceleration

In the subsection on graphing the motion of accelerated motion, it was pointed out that for constant-velocity motion, x-versus-t is linear. Since a is related to v in the same way that v is related to x, we can also see that for motion with constant acceleration, the v-versus-t plot is linear, $v = at + v_0$, its slope equals the acceleration, and the intercept equals the initial velocity.

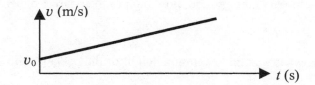

Suppose, however, that you only have position-versus-time data. The equation $x = \frac{1}{2}at^2 + v_0t + x_0$ tells us that this plot is parabolic. Is there any good way to get the acceleration from this set of data? The answer is yes if the motion starts from rest. For this special case of starting from rest, the equation for the displacement as a function of time reduces to

$$x - x_0 = \frac{1}{2}at^2$$

Notice that although Δx is quadratic in t, it is linear in t^2. So, if you treat t^2 as a single variable and plot Δx-versus-t^2, you will get a linear curve. The slope of this line equals half of the acceleration. Just double the slope and you're done.

Practice Quiz

10. Given the x-versus-t graph shown on the right, which type of motion is most likely represented
    (a) constant velocity motion
    (b) accelerated motion
    (c) motionless particle
    (d) none of the above

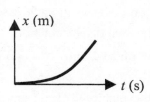

Practice Problems

1. A car is moving in such a way that its position can be described by the formula:
    $$x = 2t^3 + 1t^2 + 2t + 4 \text{ m}$$
    What is its speed, to the nearest tenth of a m/s, at $t = 2.5$?

2. In the previous problem, to the nearest of a m/s^2, what is the acceleration?

3. A car accelerates at a constant rate from zero to 38.1 m/s in 10 seconds and then slows to 14.2 m/s in 5 seconds. What is its average acceleration during the 15 seconds, to the nearest tenth of a m/s^2?

4. What was the acceleration during the first 10 seconds in the previous problem?

5. A car traveling at 15 m/s accelerates at 3.48 m/s² for 13 seconds. To the nearest meter, how far does it travel?

6. To the nearest tenth of a m/s, what is the final velocity of the car in the previous problem?

7. A passenger in a helicopter traveling upwards at 17 m/s accidentally drops a package out the window. If it takes 13 seconds to reach the ground, how high to the nearest meter was the helicopter when the package was dropped?

8. To the nearest meter, what was the maximum height of the package above the ground in the previous problem?

9. A speeder traveling at 35 m/s passes a motorcycle policeman at rest at the side of the road. The policeman accelerates at 3.49 m/s². To the nearest tenth of a second, how long does it take the policeman to catch the speeder?

10. To the nearest tenth of a meter, how far can a runner running at 11 m/s run in the time it takes a rock to fall from rest 90 meters?

Selected Solutions to End of Chapter Problems

9. **(a)** We take north as the positive y direction. Each segment of motion is at constant speed and therefore, the distance in each case is given by $\Delta y_i = v_i \, \Delta t_i$. Adding the displacements from each segment, the total displacement is

$$\Delta \vec{y} = [(30 \text{ mi/h})(2 \text{ min}) + (45 \text{ mi/h})(3 \text{ min}) + (30 \text{ mi/h})(2 \text{ min})](1 \text{ h}/ 60 \text{ min}) \hat{j} = \boxed{(4.3 \text{ mi})\hat{j}}.$$

(b) Average velocity is displacement divided by the total time. The total time for the displacement is

$$\Delta t = 2 \text{ min} + (30 \text{ s})(1 \text{ min}/ 60 \text{ s}) + 3 \text{ min} + (3 \text{ s})(1 \text{ min}/ 60 \text{ s}) + 2 \text{ min} = 7.55 \text{ min}$$

Therefore,

$$\vec{v}_{av} = \frac{\Delta \vec{y}}{\Delta t} = \frac{(4.25 \text{ mi}) \hat{j}}{7.55 \text{ min}} = (0.56 \text{ mi/min})\hat{j}$$

21. We will take the origin as the location at $t = 0$ s. Together with the fact that it starts from rest, our equation for the position as a function of time is $x = \frac{1}{2}at^2$.

At $t = 8$ s: $x_8 = \frac{1}{2}a(8\text{s})^2$.

At $t = 12$ s: $x_{12} = \frac{1}{2}a(12\text{s})^2$.

We are given that $x_{12} - x_8 = 64\,\text{m}$. Therefore,

$$\tfrac{1}{2}a(12\,\text{s})^2 - \tfrac{1}{2}a(8\,\text{s})^2 = 64\,\text{m} \quad \Rightarrow \quad a = \frac{128\,\text{m}}{(12\,\text{s})^2 - (8\,\text{s})^2} = \boxed{1.6\,\text{m/s}^2}$$

35. After falling, from rest, through a distance h, the speed of an object will be $v = (2gh)^{1/2}$. This speed is the initial speed, v_0, of the weight just before it touches the surface. The weight then falls through a distance Δx in the mud, undergoing an acceleration a, reaching a final speed $v = 0$. Using the equation $v^2 = v_0^2 + 2a\Delta x = 2gh + 2a\Delta x = 0$, we can solve for the acceleration

$$a = -gh/\Delta x = -(9.8\ \text{m/s}^2)(6\ \text{m})/(0.004\ \text{m}) = -1 \times 10^4\ \text{m/s}^2.$$

So its magnitude is $a = \boxed{1 \times 10^4\ \text{m/s}^2}$.

The average speed of the weight as it decelerates in the mud is

$$v_{\text{av}} = \tfrac{1}{2}(v_0 + v) = \tfrac{1}{2}(2gh)^{1/2} = \tfrac{1}{2}[(2(9.8\ \text{m/s}^2)(6\ \text{m})]^{1/2} = 5.42\ \text{m/s}$$

So, the time it took to stop is then

$$t = \Delta x/v_{\text{av}} = 0.004\ \text{m}/(5.42\ \text{m/s}) = \boxed{7 \times 10^{-4}\ \text{s}}.$$

59. We use a coordinate system with the origin at the ground and up as the positive y direction. We label the first rock A and the second rock B. We can find the time for rock A to hit the ground using

$$y_A = y_{0A} + v_{0A}t_A + \tfrac{1}{2}at^2$$

Noting that the final position is 0 (the ground), we get the quadratic equation

$$0 = 10\ \text{m} + (22\ \text{m/s})\,t_A + \tfrac{1}{2}(-9.8\ \text{m/s}^2)\,t_A^2,$$

which gives a positive and a negative answer; the positive answer is $t_A = 4.9$ s.

Using the same approach, the time for rock B to hit the ground is determined by

$$y_B = y_{0B} + v_{0B}t_B + \tfrac{1}{2}at^2;$$
$$0 = 10\ \text{m} + 0 + \tfrac{1}{2}(-9.8\ \text{m/s}^2)t_B^2$$

which gives $t_B = 1.4$ s. Thus rock B must be released $\Delta t = t_A - t_B = \boxed{3.5\ \text{s}}$ after rock A.

65. By definition, acceleration is the time derivative of velocity; so, the magnitude of the acceleration is the time derivative of the instantaneous speed. Using this fact, we have

$$a(t) = \frac{dv}{dt} = \frac{d}{dt}\left[4.0\,\tfrac{\text{m}}{\text{s}} + (8.0\,\tfrac{\text{m}}{\text{s}})e^{-(0.5\text{s}^{-1})t}\right] = \boxed{-(4.0\,\tfrac{\text{m}}{\text{s}})e^{-(0.5\text{s}^{-1})t}}$$

Plug this into the expression for the velocity to get

$$v(t) = 4.0\,\tfrac{\text{m}}{\text{s}} + (8.0\,\tfrac{\text{m}}{\text{s}})e^{-(0.5\text{s}^{-1})t} = 4.0\,\tfrac{\text{m}}{\text{s}} - 2a(t)$$

Solving this for acceleration then gives

$$a(v) = \boxed{2.0\,\tfrac{\text{m}}{\text{s}} - \tfrac{1}{2}v}.$$

73. The height reached is determined by the initial velocity. We assume the same initial velocity of the jumper on the Moon and Earth. With a vertical velocity of 0 m/s at the highest point, from the equation $v^2 = v_0^2 + 2ah = 0$, taking $a = -g$, we get

$$v_0^2 = 2g_E h_E = 2g_M h_M$$

Using the given fact that $g_M = g_E/6$ we have that $2g_E h_E = 2(g_E/6)h_M$. This reduces to

$$h_E = h_M/6 \quad \Rightarrow \quad h_M = 6h_E = 6(2\,\text{m}) = \boxed{12\,\text{m}}\,.$$

Answers to Practice Quiz

1. (d) **2.** (b) **3.** (c) **4.** (c) **5.** (b) **6.** (c) **7.** (b) **8.** (c) **9.** (d) **10.** (b)

Answers to Practice Problems

1. 44.5 m/s   **2.** 32.0 m/s^2

3. 0.9 m/s^2 **4.** 3.8 m/s^2

5. 489 m   **6.** 60.2 m/s

7. 607 m   **8.** 627 m

9. 20.1 s   **10.** 47.1 m

CHAPTER 3

MOTION IN TWO AND THREE DIMENSIONS

Chapter Objectives

After studying this chapter, you should

1. be able to represent position, displacement, velocity, and acceleration as two- and three-dimensional vectors.
2. be able to apply the equations for two-dimensional motion to a projectile.
3. be able to describe uniform circular motion.
4. be able to use velocity and acceleration vectors to analyze relative motion.

Chapter Review

In Chapter two, we studied motion in one dimension. In this chapter, we extend those ideas to motion in two and three dimensions (mainly two). This extension will allow us to describe a much greater variety of motions. The concepts studied here will be applied throughout your study of physics.

3–1 Position and Displacement

To specify the position of an object in a Cartesian coordinate system, we write the vector $\vec{r}$ whose x-, y-, and z-components are given by the values of the x-, y-, and z-coordinates of the point at which it is located. Generally, this object will be moving so that its position will be a function of time

$$\vec{r}(t) = x(t)\hat{i} + y(t)\hat{j} + z(t)\hat{k}$$

As before, the displacement of a particle is its change in position. So, the displacement of a particle that moves from point P to point Q is

$$\Delta\vec{r} = \vec{r}_Q - \vec{r}_P$$

Practice Quiz

1. An object moves in the x-y plane from point (1.3 m, −3.5 m) to point (−4.8 m, 5.1 m). What is its displacement (in m)?
   (a) $1.3\hat{i} - 3.5\hat{j}$ (b) $-6.1\hat{i} + 8.6\hat{j}$ (c) $-3.5\hat{i} + 1.6\hat{j}$ (d) $-4.8\hat{i} + 5.1\hat{j}$ (e) $6.1\hat{i} - 8.6\hat{j}$

3–2 Velocity and Acceleration

The generalization of velocity and acceleration to two and three dimensions follows the same pattern as the generalizations of position and displacement did.

Velocity

The average velocity is the displacement $\Delta \vec{r}$ divided by the elapsed time Δt. Recall that when dividing a vector by a scalar, we divide the scalar into each component of the vector; hence,

$$\vec{v}_{av} = \frac{\Delta \vec{r}}{\Delta t} = \frac{\Delta x}{\Delta t}\hat{i} + \frac{\Delta y}{\Delta t}\hat{j}$$

where $\Delta x = x_2 - x_1$ and $\Delta y = y_2 - y_1$.

The instantaneous velocity is the limit of the average velocity as the time interval approaches zero:

$$\vec{v} = \lim_{\Delta t \to 0} \frac{\Delta \vec{r}}{\Delta t} = \frac{d\vec{r}}{dt}$$

When taking the derivative of a vector, each component of the vector is differentiated; therefore,

$$\vec{v} = \frac{d\vec{r}}{dt} = \frac{dx}{dt}\hat{i} + \frac{dy}{dt}\hat{j}$$

This shows that the components of the instantaneous velocity are given by

$$v_x = \frac{dx}{dt} \quad \text{and} \quad v_y = \frac{dy}{dt}$$

Finally, we recall that the instantaneous speed of the particle is given by the magnitude of the instantaneous velocity. In two dimensions, this magnitude is

$$v = |\vec{v}| = \sqrt{v_x^{\,2} + v_y^{\,2}}$$

Example 3–1 Suppose you are driving northwest at 13.5 m/s for exactly half an hour when you run out of gas. Frustrated, you walk for 40.0 min to the nearest gas station, which is 2.40 km away at 30° north of west from where your car stopped. Determine your average velocity for this trip. State the answer both as a magnitude with an angular direction and using unit vectors.

Setting it up:

It is helpful to draw a diagram for the trip. The diagram on the right shows the three relevant displacement vectors. The vector $\vec{d}_c$ is your displacement while driving the car, $\vec{d}_w$ is your displacement while walking, and $\vec{r}$ is your total displacement. (For clarity, the 30° angle is actually drawn larger than 30° and the arrow on $\vec{r}$ is not at its end.) The following information is given:

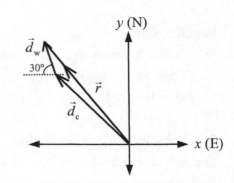

Given : $v_c = 13.5$ m/s, $t_c = 30$ min, $t_w = 40$ min, $d_w = 2.40$ km

$\qquad \theta_w = 30.0°$ N of W

Find : v_{av}, θ_{av}, $\vec{v}_{av}$

Strategy: Because average velocity is displacement divided by elapsed time, we will first calculate the displacement vectors, and then determine the average velocity.

Working it out:

The magnitude of $\vec{d}_c$ can be determined from the driving speed and time

$$d_c = v_c t_c = (13.5 \text{ m/s})(30 \text{ min})\left(\frac{60 \text{ s}}{\text{min}}\right) = 24,300 \text{ m}$$

The fact that the car was moving northwest means that the direction is

$$\theta_c = 90° + 45° = 135°$$

Therefore, the x and y components of $\vec{d}_c$ are

$$d_{cx} = d_c \cos\theta_c = (24,300 \text{ m})\cos(135°) = -17,183 \text{ m}$$
$$d_{cy} = d_c \sin\theta_c = (24,300 \text{ m})\sin(135°) = 17,183 \text{ m}$$

Similarly, the angle we wish to use for the direction of $\vec{d}_w$ is

$$\theta_w = 180° - 30° = 150°$$

The x and y components of $\vec{d}_w$ are

$$d_{wx} = d_w \cos\theta_w = (2400 \text{ m})\cos(150°) = -2,078 \text{ m}$$
$$d_{wy} = d_w \sin\theta_w = (2400 \text{ m})\sin(150°) = 1,200 \text{ m}$$

The total displacement is the sum of the two displacements, $\vec{r} = \vec{d}_c + \vec{d}_w$. Therefore, the components of $\vec{r}$ are given by

$$r_x = d_{cx} + d_{wx} = -17,183 \text{ m} - 2,078 \text{ m} = -19,261 \text{ m}$$
$$r_y = d_{cy} + d_{wy} = 17,183 \text{ m} + 1,200 \text{ m} = 18,383 \text{ m}$$

Having the components of the total displacement, we now determine the elapsed time to be

$$t = t_c + t_w = 30 \text{ min} + 40 \text{ min} = 70 \text{ min}\left(\frac{60 \text{ s}}{\text{min}}\right) = 4200 \text{ s}$$

With this information, we are now able to determine the components of the average velocity as

$$v_{av,x} = \frac{r_x}{t} = \frac{-19,261 \text{ m}}{4200 \text{ s}} = -4.586 \text{ m/s}$$

$$v_{av,y} = \frac{r_y}{t} = \frac{18,383 \text{ m}}{4200 \text{ s}} = 4.377 \text{ m/s}$$

The magnitude of the average velocity is given by

$$v_{av} = \left(v_{av,x}^2 + v_{av,y}^2\right)^{1/2} = \left[(-4.586 \text{ m/s})^2 + (4.377 \text{m/s})^2\right]^{1/2} = 6.34 \text{ m/s}$$

The reference angle for this vector is

$$\theta_{ref} = \tan^{-1}\left(\frac{v_{av,y}}{v_{av,x}}\right) = \tan^{-1}\left(\frac{4.377 \text{ m/s}}{-4.586 \text{ m/s}}\right) = -43.7°$$

From the signs of the components we can see that this angle is measured above the $-x$ axis; therefore, the magnitude and direction can be stated the following way:

$$\vec{v}_{av} \text{ is 6.34 m/s at 43.7° north of west.}$$

Since we also know the components of $\vec{v}_{av}$, it is straightforward to write it out using unit vectors,

$$\vec{v}_{av} = -(4.59 \text{ m/s})\hat{i} + (4.38 \text{ m/s})\hat{j}$$

What do you think? Is the average velocity calculated in this problem equal to the average of the velocities while driving and walking? If not, how does it relate to those two velocities?

Practice Quiz

2. Which of the following (in m/s) gives the correct average velocity of a particle that moves from position $\vec{r}_1 = 2.0\hat{i} + 5.0\hat{j}$ to position $\vec{r}_2 = 8.0\hat{i} - 3.0\hat{j}$, both in meters, in 5.0 s?

    (a) $1.2\,\hat{i} - 1.6\,\hat{j}$ (b) $6.0\,\hat{i} - 8.0\,\hat{j}$ (c) $2.0\,\hat{i} + 0.4\,\hat{j}$ (d) $30\,\hat{i} - 40\,\hat{j}$ (e) $1.6\,\hat{i} - 0.6\,\hat{j}$

Acceleration

The average acceleration is the change in velocity divided by the elapsed time

$$\vec{a}_{av} = \frac{\Delta \vec{v}}{\Delta t}$$

The change in velocity is the difference between the final and initial velocities, $\Delta \vec{v} = \vec{v}_2 - \vec{v}_1$. The instantaneous acceleration is the limit of the average acceleration as the time interval approaches zero:

$$\vec{a} = \lim_{\Delta t \to 0} \frac{\Delta \vec{v}}{\Delta t} = \frac{d\vec{v}}{dt}$$

Therefore, the acceleration vector can be written as

$$\vec{a} = \frac{dv_x}{dt}\hat{i} + \frac{dv_y}{dt}\hat{j} = \frac{d^2x}{dt^2}\hat{i} + \frac{d^2y}{dt^2}\hat{j} = a_x\hat{i} + a_y\hat{j}$$

Example 3–2 The position of a particle is given exactly by the function $\vec{r}(t) = 1.5[\cos(3t)\hat{i} + \sin(3t)\hat{j}]$, where the arguments of the sine and cosine functions are to be in radians. What is the acceleration of this particle at $t = 1.25$ s assuming that $\vec{r}(t)$ is given in SI?

Setting it up:

The function $\vec{r}(t)$ does not specify the units of the numerical quantities; this is a common practice as it is assumed that you can infer what the units must be given SI. Let us write this function showing the units explicitly. Because $\vec{r}(t)$ is a position vector, its dimension is [L] and so its SI unit is meter; therefore the overall factor of 1.5 must be in meters. Also, since we are told that the arguments of the trigonometric functions are in radians (a dimensionless quantity), and we know that t is in seconds, the 3 must be in rad/s. Thus, we are given the following information:

Given: $\vec{r}(t) = (1.5 \text{ m})\left\{\cos\left[(3\tfrac{\text{rad}}{\text{s}})t\right]\hat{i} + \sin\left[(3\tfrac{\text{rad}}{\text{s}})t\right]\hat{j}\right\}$;   **Find**: $\vec{a}(t)$

Strategy: We know that acceleration is the second time derivative of position. So, to determine $\vec{a}(t)$, we must take two derivatives of the given function for $\vec{r}(t)$.

Working it out:

The first derivative of position gives the velocity as a function of time

$$\vec{v}(t) = \frac{d\vec{r}}{dt} = \left(3\tfrac{\text{rad}}{\text{s}} \times 1.5\ \text{m}\right)\left\{-\sin\left[(3\tfrac{\text{rad}}{\text{s}})t\right]\hat{i} + \cos\left[(3\tfrac{\text{rad}}{\text{s}})t\right]\hat{j}\right\} = \left(4.5\ \text{m/s}\right)\left\{-\sin\left[(3\tfrac{\text{rad}}{\text{s}})t\right]\hat{i} + \cos\left[(3\tfrac{\text{rad}}{\text{s}})t\right]\hat{j}\right\}$$

Notice the result $3\tfrac{\text{rad}}{\text{s}} \times 1.5\ \text{m} = 4.5\ \text{m/s}$. The units work out this way because angle is a dimensionless quantity. Taking the derivative of the velocity gives

$$\vec{a}(t) = \frac{d\vec{v}}{dt} = \left(3\tfrac{\text{rad}}{\text{s}} \times 4.5\ \text{m/s}\right)\left\{-\cos\left[(3\tfrac{\text{rad}}{\text{s}})t\right]\hat{i} - \sin\left[(3\tfrac{\text{rad}}{\text{s}})t\right]\hat{j}\right\} = -\left(13.5\ \text{m/s}^2\right)\left\{\cos\left[(3\tfrac{\text{rad}}{\text{s}})t\right]\hat{i} + \sin\left[(3\tfrac{\text{rad}}{\text{s}})t\right]\hat{j}\right\}$$

Now, we only need to evaluate this acceleration function at $t = 1.25$ s.

$$\vec{a}(1.25) = -\left(13.5\ \text{m/s}^2\right)\left\{\cos\left[(3\tfrac{\text{rad}}{\text{s}})(1.25\,\text{s})\right]\hat{i} + \sin\left[(3\tfrac{\text{rad}}{\text{s}})(1.25\,\text{s})\right]\hat{j}\right\}$$

$$= -\left(13.5\ \text{m/s}^2\right)\left\{-0.8206\hat{i} - 0.5716\hat{j}\right\}$$

$$= \left(11.1\ \text{m/s}^2\right)\hat{i} + \left(7.72\ \text{m/s}^2\right)\hat{j}$$

What do you think? What kind of motion does the given function describe? [Hint: How does the magnitude of the position vector depend on time?]

Practice Quiz

3. After accelerating for 6.75 s, an object has an average acceleration of $\vec{a}_{av} = 8.21\hat{i} + 1.71\hat{j}$ in m/s^2. What was its change in velocity? (Answers are in m/s.)

    (a) $187\,\hat{i} + 39.0\,\hat{j}$ (b) $1.22\,\hat{i} + 0.253\,\hat{j}$ (c) $55.4\,\hat{i} + 11.5\,\hat{j}$ (d) $4.11\,\hat{i} - 0.855\,\hat{j}$ (e) 0

4. The velocity of an object is given by $v = 4e^{-2t}$ in SI. What is the acceleration of this object as a function of time?

    (a) $a = \left(-2\ \text{m/s}^2\right)e^{-(2/s)t}$ (b) $a = \left(4\ \text{m/s}\right)e^{-(2/s)t}$ (c) $a = \left(2\ \text{m/s}\right)e^{-t/2}$

    (d) $a = \left(16\ \text{m/s}^2\right)e^{-2t/s}$ (e) $a = \left(-8\ \text{m/s}^2\right)e^{-(2/s)t}$

Trajectories

The path through which an object moves is called its **trajectory**. There are two important things to remember about representing trajectories. The first thing is that the trajectory itself is mapped out by the tip of the position vector, $\vec{r}$, as time passes. The second thing to remember is that the instantaneous velocity of the particle is always tangent to the trajectory.

3–3—3–4 Motion with Constant Acceleration and Projectile Motion

In two dimensions, motion with constant acceleration is restricted to a plane which we take to be the xy-plane. Describing this type of motion in two dimensions is equivalent to describing two independent one-dimensional motions, one motion being in the x-direction (i.e., for the x-components of displacement, velocity, and acceleration), and the other being in the y-direction. As a result of this fact, we can use the same four equations from our study of constant acceleration in one dimension for each component of the two-dimensional motion. (You may also add the fifth equation from the Tips section of Chapter two.)

x-direction of Motion	y-direction of Motion
$v_x = v_{0x} + a_x t$	$v_y = v_{0y} + a_y t$
$v_{av,x} = \frac{1}{2}\left(v_{0x} + v_x\right)t$	$v_{av,y} = \frac{1}{2}\left(v_{0y} + v_y\right)t$
$v_x^2 = v_{0x}^2 + 2a_x\left(x - x_0\right)$	$v_y^2 = v_{0y}^2 + 2a_y\left(y - y_0\right)$
$x = x_0 + v_{0x}t + \frac{1}{2}a_x t^2$	$y = y_0 + v_{0y}t + \frac{1}{2}a_y t^2$

Our most common application of constant acceleration in two dimensions is that of **projectile motion**. This motion is that of an object projected with some initial velocity and then allowed to fall freely. Effects such as air resistance and Earth's rotation, that sometimes causes an object's motion to differ significantly from that of pure free fall, are ignored. The key point for understanding projectile motion is that the acceleration due to gravity, $\vec{g}$, acts only vertically downward, and since gravity is the only influence, there is no acceleration in the horizontal direction. This fact means that **although there is gravitational acceleration vertically, the horizontal motion is that of constant velocity**.

It is convenient to adopt a standard coordinate system for projectile motion. This standard takes the positive direction for vertical motion to be upward, making downward the negative direction. This means that the acceleration of gravity points in the negative y-direction. For the horizontal motion, the positive direction is to the right, and the negative direction to the left. The origin of this standard coordinate system is located at the starting point of the projectile. Given this coordinate system, we can rewrite the equations for two-dimensional motion in a form specific to projectile motion. To accomplish this rewriting, we make use of several facts:

$$x_0 = 0, \quad a_x = 0, \quad v_x = v_{0x} = v_0 \cos\theta_0$$
$$y_0 = 0, \quad a_y = -g, \quad v_{0y} = v_0 \sin\theta_0$$

where θ_0 is the **elevation angle**, which is the angle of the initial velocity as measured from the horizontal direction. With these substitutions, the equations of projectile motion are as follows:

Horizontal Motion	Vertical Motion
$v_x = v_0 \cos\theta_0$	$v_y = v_0 \sin\theta_0 - gt$
$x = \left(v_0 \cos\theta_0\right)t$	$v_{av,y} = \frac{1}{2}\left(v_0 \sin\theta_0 + v_y\right)t$
	$v_y^2 = v_0^2 \sin^2\theta_0 - 2gy$
	$y = \left(v_0 \sin\theta_0\right)t - \frac{1}{2}g t^2$

This set of equations describes projectile motion for both zero and nonzero elevation angles.

Example 3–3 A person stands 12.5 m from the base of a building that is 40.0 m tall. The person wants to toss a wrench to his coworker who is working on the roof. If he releases the tool at a height of 1.00 m above the ground, what speed and direction must he give the tool if it is to just make it onto the roof?

Setting it up:

It is helpful to draw a diagram for this problem. Our sketch shows the wrench being tossed so that it just barely makes it onto the roof. With the origin chosen to be the point of launch, the given information reduces to the following

Given: $x_0 = 0$, $y_0 = 0$, $x = 12.5$ m, $y = 39.0$ m

Find: v_0, θ_0

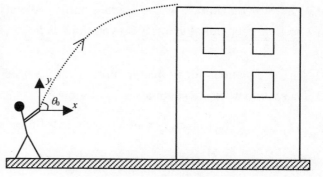

Strategy: Because the wrench just barely makes it onto the roof, the top of the roof is the maximum height of the wrench, where $v_y = 0$. Treating the motions separately, we can find the components of v_0. From these values we can determine the magnitude and direction.

Working it out: We first want the components of the initial velocity. Recalling that $v_{0y} = v_0 \sin\theta_0$, we can examine the equations for projectile motion to see if one of them can be used to solve for v_{0y} without any intermediate calculations. We are not given any information about time, so looking at the expression that does not require us to know t, we see that we are, in fact, able to solve for v_{0y}

$$v_y^{\ 2} = v_0^2 \sin^2\theta_0 - 2gy \quad \Rightarrow \quad 0 = v_{0y}^2 - 2gy$$

So that,

$$v_{0y} = \sqrt{2gy} = \sqrt{2\left(9.80 \text{ m/s}^2\right)\left(39.0 \text{ m}\right)} = 27.65 \text{ m/s}$$

To get the x-component of the initial velocity, $v_{0x} = v_x = v_0 \cos\theta_0$, we note that the only equation involving v_{0x}, $x = v_{0x}\, t$, requires the time t. Thus, we seek a vertical equation involving t with all other quantities being known. Doing so, we find

$$v_y = v_0 \sin\theta_0 - gt \quad \Rightarrow \quad 0 = v_{0y} - gt$$

This gives

$$t = \frac{v_{0y}}{g} = \frac{27.65 \text{ m/s}}{9.80 \text{ m/s}^2} = 2.821 \text{ s}$$

Now, we can solve for v_{0x}

$$v_{0x} = \frac{x}{t} = \frac{12.5 \text{ m}}{2.821 \text{ s}} = 4.431 \text{ m/s}$$

The speed that he must impart to the tool is given by the magnitude of the initial velocity

$$v_0 = \sqrt{v_{0x}^2 + v_{0y}^2} = \sqrt{(4.431 \text{ m/s})^2 + (27.65 \text{ m/s})^2} = 28.0 \text{ m/s}$$

The initial direction is determined by

$$\theta_0 = \tan^{-1}\left(\frac{v_{0y}}{v_{0x}}\right) = \tan^{-1}\left(\frac{27.65 \text{ m/s}}{4.431 \text{ m/s}}\right) = 80.9°$$

What do you think? What is the initial velocity vector in unit-vector notation? Which form of specifying the initial velocity do you think is more useful in everyday situations?

The equations for projectile motion can be used to derive several important properties of the motion. These properties are summarized as follows:

* The trajectory that a projectile follows is a parabola described by the equation.

$$y = (\tan\theta_0)x - \left(\frac{g}{2v_0^2 \cos^2\theta_0}\right)x^2$$

* If the initial and final elevations are the same, the **range** (R) of a projectile, which is the horizontal distance it travels before landing, is given by

$$R = \frac{v_0^2}{g}\sin 2\theta_0$$

The launch angle that produces the maximum range is $\theta_0 = 45°$.

* The amount of time a projectile spends in the air is sometimes called the **flight time** T. If the initial and final elevations are the same, this time is given by

$$T = \frac{2v_0}{g}\sin\theta_0$$

* If the initial and final elevations are the same, the time it takes a projectile launched at some upward angle to reach its maximum height equals the time it takes to fall back down from its maximum height.

* The maximum height, h, that a projectile will reach above its initial height is

$$h = \frac{(v_0 \sin\theta)^2}{2g}$$

* The speed that a projectile has at a given height on its way up is equal to the speed it will have at that same height on its way back down.

* At a given height, the angle of the velocity of a projectile above the horizontal on its way up equals the angle of the velocity below the horizontal on its way down.

Notice that all these characteristics are determined by the initial velocity given to the projectile.

Example 3–4 A golf ball sitting on level ground is struck and given an initial velocity of 41.2 m/s at an angle of 58.0°. **(a)** How high does the ball go into the air? **(b)** How far does it travel? **(c)** How long is the ball in the air?

Setting it up: The following information is given in the problem:

Given: $v_0 = 41.2$ m/s, $\theta = 58.0°$;   **Find**: **(a)** h,   **(b)** R,   **(c)** T

Strategy: We are given the initial velocity which completely specifies the motion. So, we can use previously derived equations to directly solve for the desired information.

Working it out:

(a)   For the maximum height we have

$$h = \frac{(v_0 \sin\theta)^2}{2g} = \frac{\left[(41.2 \text{ m/s})\sin(58.0°)\right]^2}{2(9.80 \text{ m/s}^2)} = 62.3 \text{ m}$$

(b)   For the range we have

$$R = \frac{2v_0^2}{g}\sin 2\theta = \frac{(41.2 \text{ m/s})^2}{9.80 \text{ m/s}^2}\sin(116°) = 156 \text{ m}$$

(c)   For the flight time we have

$$T = \frac{2v_0}{g}\sin\theta = \frac{2(41.2 \text{ m/s})}{9.80 \text{ m/s}^2}\sin(58.0°) = 7.13 \text{ s}$$

What do you think? Can you use the list of properties to determine the speed of the golf ball when it strikes the ground?

Practice Quiz

5.   A stone is thrown horizontally from the top of a 50.0-m building with a speed of 12.3 m/s. How far from the building will it land?

   (a) 30.8 m   **(b)** 39.3 m   **(c)** 3.19 m   **(d)** 615 m   **(e)** 50.0 m

6.   A projectile is launched from the ground with a speed of 23.7 m/s at 33.0° above the horizontal. What is its speed when it is at its maximum height above the ground?

   (a) 15.4 m/s   **(b)** 0.00 m/s   **(c)** 12.9 m/s   **(d)** 23.7 m/s   **(e)** 19.9 m/s

7.   A projectile is launched from the ground with a speed of 53.4 m/s at 68.0° above the horizontal. Take its initial position as the origin, what are its (x, y) coordinates after 8.24 seconds?

   (a) (440 m, 440 m)   **(b)** (440 m, 107 m)   **(c)** (408 m, 333 m)   **(d)** (165 m, 75.3 m)

3–5 Uniform Circular Motion

Motion doesn't always take place in straight-line paths. Therefore, we need to study how to handle motion along curves. Specifically, this section focuses on circular motion, or at least motion along a circular section. If the object that moves along this circular path does so at constant speed, we called it **uniform circular motion**.

It is more convenient to analyze this type of motion using plane polar coordinates rather than Cartesian coordinates. As the figure below shows, in **plane polar coordinates**, the two variables that specify the position of a particle are r and ϕ. The variable r specifies the distance of the particle from the origin; and ϕ specifies the angle (usually relative to the positive x-direction) at which the particle can be found.

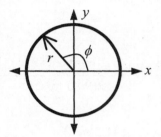

For the case of uniform circular motion, r has a constant value equal to the radius of the circular path $r = R$, while ϕ takes on values from 0 to 2π.

The uniform speed that characterizes this motion is given by the rate at which the particle moves through arc length around this path

$$v = R\frac{d\phi}{dt} = R\omega$$

where $\omega = d\phi/dt$ is the rate at which the angle ϕ changes called the **angular speed**. The time it takes the object to move once around the circular path is called the **period** of the motion T. The period is related to v and ω by

$$T = \frac{2\pi R}{v} = \frac{2\pi}{\omega}$$

The inverse of the period specifies the number of revolutions that an object makes per unit time and is called the **frequency** of the motion f

$$f = \frac{1}{T}$$

In SI, frequencies are quoted in the unit hertz (Hz) instead of s^{-1}. There is also an often used relationship between the angular speed and the frequency

$$\omega = 2\pi f$$

For uniform circular motion, the position, velocity, and acceleration vectors have certain relationships with each other. The position vector $\vec{r}$ has magnitude equal to the value of the r coordinate in plane polar coordinates $|\vec{r}| = r$. We can specify the direction of the position vector by defining a unit vector

$$\hat{r} \equiv \frac{\vec{r}}{r}$$

so that $\vec{r} = r\hat{r}$. The velocity vector has a magnitude equal to the uniform speed v, and is always directed tangent to the circular path as mentioned in section 3–2. Because only the direction of the velocity changes, the acceleration in uniform circular motion must be such that the object neither speeds up nor slows down during its motion. This situation can only occur if the acceleration is always perpendicular to the velocity. Analysis of this type of motion shows that the acceleration always points toward the center of the circular path; for this reason it is called a **centripetal acceleration**. The acceleration is given by

$$\vec{a} = -\frac{v^2}{r}\hat{r}$$

Notice that the position, velocity, and acceleration vectors all have constant magnitudes, but their directions vary continually.

Example 3–5 (a) Show that the position vector given in Example 3–2 describes uniform circular motion. **(b)** Determine the magnitude of the centripetal acceleration of this motion. **(c)** Calculate the angular speed, period, and frequency of this motion.

Setting it up: From Example 3–2 we know the following information:

Given: $\vec{r}(t) = (1.5 \text{ m})\left\{\cos\left[(3\tfrac{\text{rad}}{\text{s}})t\right]\hat{i} + \sin\left[(3\tfrac{\text{rad}}{\text{s}})t\right]\hat{j}\right\}$, $\vec{v}(t) = (4.5 \text{ m/s})\left\{-\sin\left[(3\tfrac{\text{rad}}{\text{s}})t\right]\hat{i} + \cos\left[(3\tfrac{\text{rad}}{\text{s}})t\right]\hat{j}\right\}$;

Find: **(a)** Is it uniform circular motion? **(b)** a **(c)** ω, T, f

Strategy: **(a)** Uniform circular motion means that the motion is at constant speed in a circular path. So, we need to determine $r(t)$ and $v(t)$ to see if the motion meets these requirements. **(b)** We can use the acceleration vector calculated in Example 3–2 and find its magnitude. **(c)** We can use the expression relating the v and ω to find ω, which then will allow us to calculate T and f.

Working it out:

(a) The magnitude of the position vector is

$$r(t) = (1.5 \text{ m})\left\{\cos^2\left[(3\tfrac{\text{rad}}{\text{s}})t\right] + \sin^2\left[(3\tfrac{\text{rad}}{\text{s}})t\right]\right\}^{1/2} = 1.5 \text{ m}$$

Clearly, this result is independent of time; so the particle is always a fixed distance from the origin. It is either stationary or moving in a circular path. The magnitude of the velocity is

$$v(t) = (4.5 \text{ m/s})\left\{\left(-\sin\left[(3\tfrac{\text{rad}}{\text{s}})t\right]\right)^2 + \cos^2\left[(3\tfrac{\text{rad}}{\text{s}})t\right]\right\}^{1/2} = 4.5 \text{ m/s}$$

Again, this result is independent of time and the particle is not stationary. Given these two facts, we can conclude that the motion is uniform circular motion with $R = 1.5$ m and $v = 4.5$ m/s.

(b) Knowing both the speed and the radius we can determine the centripetal acceleration

$$a = \frac{v^2}{R} = \frac{(4.5 \text{ m/s})^2}{1.5 \text{ m}} = 13.5 \text{ m/s}^2$$

(c) The angular speed is related to the speed according to $v = R\omega$. Thus,

$$\omega = \frac{v}{R} = \frac{4.5 \text{ m/s}}{1.5 \text{ m}} = 3 \text{ s}^{-1} = 3 \text{ rad/s}$$

Now that we know the angular speed, we can quickly find the frequency using $\omega = 2\pi f$. Thus,

$$f = \frac{\omega}{2\pi} = \frac{3 \text{ s}^{-1}}{2\pi} = 0.477 \text{ s}^{-1} = 0.477 \text{ Hz}$$

From the frequency, we find the period to be

$$T = \frac{1}{f} = \frac{1}{0.4775 \text{ s}^{-1}} = 2.09 \text{ s}$$

What do you think? Comparing the results of parts (a) and (c) to the expression for $\vec{r}(t)$, how would you rewrite $\vec{r}(t)$, substituting variables for the appropriate numerical quantities?

Practice Quiz

8. For an object undergoing circular motion at constant speed,
    (a) the velocity is constant
    (b) the acceleration is constant
    (c) the direction of the velocity is constant
    (d) the direction of the acceleration is constant
    (e) none of the above

9. Which of the following correctly identifies the relationship between the directions of velocity and acceleration for objects in circular motion at constant speed?
    (a) They are in the same direction.
    (b) They are in opposite directions.
    (c) They are perpendicular to each other.
    (d) Their directions may differ by any angle depending on the curve of the arc.
    (e) none of the above

3–6 Relative Motion

The motion of any object can be, and often is, observed by different people. Let us refer to anything, whether a person or an experimental device, that can detect the motion of an object as an **observer**. Every observer makes his or her (or its) observations from a particular **frame of reference** (or reference frame). We can think of a frame of reference as a coordinate system that moves along with the observer. Two observers that are not at rest with respect to each other are in **relative motion**. The question being addressed here is, how do the observations of observers in relative motion compare to each other?

Consider the following diagram, which shows two coordinate systems (reference frames), A and B, and an object at point P that is in motion relative to each coordinate system.

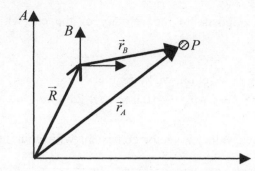

From the diagram, we can see that the position of P as measured by observers in A and B is related by the vector equation

$$\vec{r}_A = \vec{R} + \vec{r}_B$$

The relationship between the velocities is obtained by taking the time derivative of the above equation

$$\vec{v}_A = \vec{u} + \vec{v}_B$$

where $\vec{u}$ is the velocity of frame B with respect to frame A. (Note that the velocity of frame A with respect to frame B is $-\vec{u}$). If $\vec{u} = $ constant, then differentiating the equation for the velocities shows that observers in frames A and B measure the same acceleration for the object at P

$$\vec{a}_A = \vec{a}_B$$

However, if $\vec{u} \neq$ constant, then the two frames will measure different accelerations

$$\vec{a}_A = \frac{d\vec{u}}{dt} + \vec{a}_B$$

Example 3–6 Tom and Jan are standing at the side of the street waiting to cross. A car is coming at 10.0 m/s. Jan decides to go ahead and run directly across the street at 2.80 m/s, whereas Tom chooses to wait. While Jan is running across, what is her velocity as measured by the driver of the car?

Setting it up:

Our sketch shows Jan, Tom, and the car, as well as the given velocity vectors. Because we seek Jan's velocity, she plays the role of particle P in the previous discussion, while the car and Tom play the roles of frames A and B respectively. Notice that the given velocities are with respect to Tom.

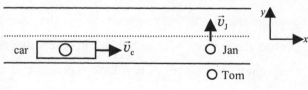

The following information is given in the problem:

Given: $\vec{v}_{cT} = 10.0$ m/s $\hat{i}$, $\vec{v}_{JT} = 2.80$ m/s $\hat{j}$;   **Find:** $\vec{v}_{Jc}$

Strategy: We need to identify the relative velocity between the reference frames $\vec{u}$. In our expression for velocity we have $\vec{v}_A = \vec{u} + \vec{v}_B$, where $\vec{u}$ is the velocity of B with respect to A. Since Tom plays the role of frame B here, then we need the velocity of Tom with respect to the car. The velocity of Tom with respect to the car is equal and opposite to the velocity of the car with respect to Tom; therefore, $\vec{u} = -10.0 \text{ m/s } \hat{i}$.

Working it out: With everything now set up, we know that

$$\vec{v}_{Jc} = \vec{u} + \vec{v}_{JT} = -10.0 \text{ m/s } \hat{i} + 2.80 \text{ m/s } \hat{j}$$

What do you think? Is the above velocity vector consistent with your intuition as to how Jan should move with respect to the car?

_____

Practice Quiz

10. If $\vec{v}_A$ is the velocity of object P with respect to frame A, and $\vec{v}_B$ is the velocity of object P with respect to frame B, which of the following equals the velocity of frame B with respect to frame A?

   (a) $-\vec{v}_A - \vec{v}_B$ **(b)** $\vec{v}_A + \vec{v}_B$ **(c)** $-\vec{v}_A + \vec{v}_B$ **(d)** $-\vec{v}_B - \vec{v}_A$ **(e)** $-\vec{v}_B + \vec{v}_A$

Reference Tools and Resources

I. Key Terms and Phrases

trajectory the path of an object in space

projectile motion the motion of an object that is projected with an initial velocity and then moves under the influence of gravity only

elevation angle the angle of the initial velocity of a projectile measured relative to the horizontal

range of a projectile the horizontal distance traveled by a projectile before it lands

flight time of a projectile the total amount of time a projectile is in the air

uniform circular motion motion at constant speed in a circular path

plane polar coordinates a two-dimensional coordinate system that specifies the location of a point using radial distance r and an angle ϕ as the coordinates

angular speed the rate at which the angular coordinate of a particle changes

period the amount of time it takes an object moving in circular motion to move once around its path

frequency the number of revolutions per second of a particle moving in circular motion

centripetal acceleration the center seeking acceleration of an object moving in uniform circular motion

observer anything that can detect the motion of an object

frame of reference a coordinate system relative to which motions can be observed and described

relative motion the motion between observers that are not at rest with respect to each other

II. Important Equations

Name/Topic	Equation	Explanation
Position	$\vec{r}(t) = x(t)\hat{i} + y(t)\hat{j} + z(t)\hat{k}$	A general position vector as a function of time.
Instantaneous velocity	$\vec{v} = \dfrac{d\vec{r}}{dt}$	The general expression for instantaneous velocity.
Instantaneous acceleration	$\vec{a} = \dfrac{d\vec{v}}{dt}$	The general expression for instantaneous acceleration.
Range of a projectile	$R = \dfrac{v_0^2}{g}\sin 2\theta_0$	The horizontal distance traveled by a projectile if the initial height equals the final height.
Flight time	$T = \dfrac{2v_0}{g}\sin\theta_0$	The total time in the air if the initial height equals the final height.
Maximum height	$h = \dfrac{(v_0\sin\theta)^2}{2g}$	The maximum height above the initial height of launch.
Uniform circular motion	$v = R\omega$	The relationship between the speed v and angular speed ω of a particle in uniform circular motion.
Frequency	$f = \dfrac{1}{T} = \dfrac{\omega}{2\pi}$	The relationship between the frequency, period, and angular speed of a particle in uniform circular motion.
Centripetal acceleration	$\vec{a} = -\dfrac{v^2}{r}\hat{r}$	The acceleration of a particle in uniform circular motion.
Relative motion	$\vec{v}_A = \vec{u} + \vec{v}_B$	The relationship between velocities measured in two different reference frames A and B.

III. Know Your Units

Quantity	Dimension	SI Unit
Angular speed (ω)	$[T^{-1}]$	rad/s
Period (T)	$[T]$	s
Frequency (f)	$[T^{-1}]$	s^{-1}

IV. Tips

When faced with situations involving relative motion, there is a mnemonic for writing down the correct vector equations relating the relative quantities (such as the velocities). This mnemonic device uses a system of subscripts in which the first subscript refers to the object whose velocity is being measured, and the second subscript refers to the observer (or frame of reference) making the measurement. For example, a velocity labeled $\vec{v}_{PA}$ refers to the velocity *of* object P *relative to* (or as measured in) the reference frame of object A. If you reverse the subscripts to get the velocity of A relative to P, this velocity, $\vec{v}_{AP}$, relates to $\vec{v}_{PA}$ by a minus sign: $\vec{v}_{AP} = -\vec{v}_{PA}$.

To see how this system of subscripts is used, consider the situation described in section 3–6 above in which a particle at point P is moving relative to two different reference frames A and B. The velocity of frame B with respect to frame A, called $\vec{u}$ in the previous discussion, would be labeled as $\vec{v}_{BA}$. Thus, the velocity of the particle at point P as measure by observers in frame A can be written as

$$\vec{v}_{PA} = \vec{v}_{PB} + \vec{v}_{BA}$$

This equation is identical to the equation $\vec{v}_A = \vec{u} + \vec{v}_B$ written previously, but notice how the subscripts relate to each other. The resultant velocity $\vec{v}_{PA}$ has P as the leftmost subscript and A as the rightmost subscript. Now, the subscripts on the two velocities whose addition produces the resultant are ordered *PBBA*. In this ordering, P is the leftmost subscript and A is the rightmost, just as with the resultant. The subscript for the system not referenced by the resultant (B in this case) is repeated in the middle. This mnemonic can be used to help analyze almost any relative motion situation as long as you stay consistent with how you label the quantities. You should be careful, however, not to fall into the trap of using this mnemonic as a substitute for understanding the physical situation being described.

Practice Problems

1. An object is traveling according to the equations

$$x = 7t^2 + 2t$$

$$y = 3t^3$$

To the nearest tenth of a meter, how far away is the object from the origin at $t = 1.6$?

2. To one decimal place, what is the speed of the object in the previous problem at the given time?

3. A ball is moving in a circle with a radius of 15 meters. It has a period of 36 seconds. To the nearest tenth of a m/s, what is its speed?

4. A ball is moving in a vertical circle with a radius of 0.96 meters. Its speed is 5.2 m/s at the top. To the nearest tenth of a m/s^2, what is its acceleration at the top?

5. Vector $\vec{A}$ is 10 units long and is at an angle of 70 degrees counterclockwise from east. Vector $\vec{B}$ is 9 units long and is at an angle of –30 degrees. What is the direction of $\vec{A} + 7.9\vec{B}$, to the nearest tenth of a degree? (Report your results in the format –180° to 180°.)

6. In the previous problem, what is the magnitude of the resultant, to the nearest tenth of a unit?

7. A car traveling at 6.3 m/s northwest reaches a curve; 7.6 seconds later it is heading north at 18.8 m/s. What is the magnitude of its average acceleration, to two decimal places?

8. In Problem 7, what is the direction of the average acceleration to the nearest tenth of a degree? (Report your results in the format –180° to 180°.)

9. A boat is traveling with a speed 5.4 mph over the water and at a direction of 154 degrees counterclockwise from east. The river itself is flowing east at 5.4 mph relative to its bank. What is the speed of the boat relative to the riverbank, to the nearest hundredth of a mph?

10. In the previous problem, to the nearest tenth of a degree, what is the angle of progress of the boat with respect to the river bank? (Report your results in the format –180° to 180°.)

Selected Solutions to End of Chapter Problems

7. As the graph suggests, we should analyze this motion in two different segments. The graph tells us what the segments are in terms of position, but, because we seek the position as a function of time, we will need to separate the segments in terms of time. Once we know the velocity in each time segment, the position is given by $\vec{r} = \vec{r}_0 + \vec{v}t$.

    For both segments, the speed is $v = (25.2 \ \text{km/h})(10^3 \ \text{m/km})(1 \ \text{h/3600 s}) = 7.00$ m/s.

    We find the orientation of the first segment from
    $\tan \theta = y_1/x_1 = (250 \ \text{m})/(150 \ \text{m}) = 1.67$;
    So, $\theta = \tan^{-1}(1.67) = 59°$.

    For the first segment, the velocity components are
    $v_{x1} = v \cos \theta = (7.00 \ \text{m/s}) \cos 59° = 3.60$ m/s.
    $v_{y1} = v \sin \theta = (7.00 \ \text{m/s}) \sin 59° = 6.00$ m/s.

The time when the turning point is reached, the end of the first segment, is
$t_1 = x_1/v_{x1} = (150\text{ m})/(3.60\text{ m/s}) = 41.6\text{ s.}$

For the second segment, the velocity components are
$v_{x2} = v = 7.00\text{ m/s;}$
$v_{y2} = 0.$

The elapsed time for the second segment is
$\Delta t = \Delta x/v_{x2} = (400\text{ m} - 150\text{ m})/(7.00\text{ m/s}) = 35.7\text{ s.}$
So, the second segment ends at
$t_2 = 41.6\text{ s} + 35.7\text{ s} = 77.3\text{ s.}$

The position vector then is given by
$$\vec{r} = \begin{cases} \vec{r}_0 + \vec{v}_1 t & (\text{for } 0 \le t \le t_1) \\ \vec{r}(t_1) + \vec{v}_2(t - t_1) & (\text{for } t_1 < t \le t_2) \end{cases}$$

For the first segment $\vec{r}_0 = 0$ and for the second segment we have
$$\vec{r} = (150\text{m})\hat{i} + (250\text{m})\hat{j} + \left[+ (7.00\text{m/s})(t - 41.6\text{ s}) \right]\hat{i}$$
$$= \left[-141\text{ m} + (7.00\text{ m/s})t \right]\hat{i} + (250\text{m})\hat{j}$$

Therefore, the position vector for the entire motion is
$$\vec{r} = \begin{cases} \left[(3.60\text{m/s})\hat{i} + (6.00\text{m/s})\hat{j} \right]t & (\text{for } 0\text{ s} \le t \le 41.6\text{ s}) \\ \left[-141\text{ m} + (7.00\text{ m/s})t \right]\hat{i} + (250\text{m})\hat{j} & (\text{for } 41.6\text{ s} < t \le 77.3\text{ s}) \end{cases}$$

23. We are given $h = H - ut - (u/B)\,e^{-Bt}$.
Because the exponent $-Bt$ must be dimensionless, $[B] = [t]^{-1} = \boxed{[T^{-1}]}$.

We get the velocity from $v = dh/dt = -u - (u/B)\,e^{-Bt}(-B) = u[-1 + e^{-Bt}]$.
Evaluating this at $t = 0$ gives $v(0) = v_0 = \boxed{0}$.
Further evaluating as $t \to \infty$ gives $v_\infty \to \boxed{-u}$.

We get the acceleration from $a = dv/dt = u[e^{-Bt}(-B)] = -Bu\,e^{-Bt}$.
At $t = 0$, this gives $a(0) = a_0 = \boxed{-Bu}$.
As $t \to \infty$ we get $a \to \boxed{0}$.

33. We choose a coordinate system with the origin at the release point, with x horizontal and y vertical.
The horizontal motion will have a constant velocity of
$$v_{0x} = v_0 \cos\theta = (225\text{ m/s}) \cos 34° = 187\text{ m/s.}$$
At the highest point, the vertical velocity will be zero, so the full speed is given by the x component
$$v = v_{0x} = \boxed{187\text{ m/s.}}$$

45. (a) The ball passes the goal posts when it has traveled the horizontal distance of 35 m

$$x = v_{0x}t \quad \Rightarrow \quad t = \frac{x}{v_{0x}} = \frac{35 \text{ m}}{(30 \text{ m/s})\cos(32°)} = \boxed{1.4\text{s}}$$

(b) To see if the kick is successful, we must find the height of the ball at this time. The equation for the y component of position is

$$y = y_0 + v_{0y}t + \tfrac{1}{2}a_yt^2 = 0 + (30 \text{ m/s})\sin 32° \, (1.38 \text{ s}) + \tfrac{1}{2}(-9.8 \text{ m/s}^2)(1.38 \text{ s})^2 = 12.6 \text{ m}.$$

Because the bar is only 4.0 m high, then $\boxed{\text{yes}}$, the kick is successful and clears the bar by 12.6 m − 4.0 m = $\boxed{8.6 \text{ m}}$.

55. Converting the speed to SI gives $v = (65 \text{ mi/h})(1.61 \times 10^3 \text{ m/mi})(1 \text{ h}/3600 \text{ s}) = 29.1 \text{ m/s}$. We require that the centripetal acceleration not exceed $0.1g$. So,

$$a = v^2/R \leq 0.1g.$$

This implies,

$$R \geq \frac{v^2}{0.1g} = \frac{(29.1 \text{ m/s})^2}{0.1(9.8 \text{ m/s}^2)} = 860 \text{ m} = 0.86 \text{ km}$$

67. During the first leg, in order to fly due south, the airplane must head southwest, as in the diagram. From the diagram we see that $\vec{v}_1 = \vec{v}_p + \vec{v}_w$. Where $\vec{v}_p$ is the velocity of the airplane with respect to the air, $\vec{v}_1$ is the velocity of the airplane with respect to the ground, and $\vec{v}_w$ is the velocity of the wind with respect to the ground.

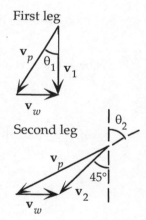

Because $\vec{v}_1$ has no x component, we know that $v_{px} = -v_{wx} = -120$ km/h. Using this result together with the facts that $v_p = 900$ km/h and that it is generally directed southwest, we can determine that

$$v_{py} = -\sqrt{v_p^2 - v_{px}^2} = -\sqrt{\left(900 \tfrac{\text{km}}{\text{h}}\right)^2 - \left(120 \tfrac{\text{km}}{\text{h}}\right)^2} = -892 \text{ km/h}$$

Again, since the x components of $\vec{v}_p$ and $\vec{v}_w$ cancel, we can say that

$$\vec{v}_1 = -892 \tfrac{\text{km}}{\text{h}} \hat{j} \quad \Rightarrow \quad v_1 = 892 \tfrac{\text{km}}{\text{h}}.$$

The distance traveled in the first leg then is

$$d_1 = (892 \text{ km/h})(2.0 \text{ h}) = 1784 \text{ km}.$$

During the second leg, the airplane must turn more toward the west. If $\vec{v}_2$ is the new velocity of the airplane with respect to the ground, then now $\vec{v}_2 = \vec{v}_p + \vec{v}_w$. The two equations for the components of $\vec{v}_2$ can be obtained from the diagram.

x-components: $(900 \text{ km/h}) \sin \theta_2 - 120 \text{ km/h} = v_2 \sin 45°$

y-components: $(900 \text{ km/h}) \cos \theta_2 = v_2 \cos 45°$.

Because sin 45° = cos 45°, we can write this as

$$(900 \text{ km/h}) \sin \theta_2 - 120 \text{ km/h} = (900 \text{ km/h}) \cos \theta_2 ,$$

which reduces to

$$\sin \theta_2 = \cos \theta_2 + 0.133.$$

By squaring both sides, we get

$$1 - \cos^2 \theta_2 = \cos^2 \theta_2 + 0.266 \cos \theta_2 + (0.133)^2,$$

which has the solution

$$\cos \theta_2 = 0.637 \text{ or } \theta_2 = 50.4°.$$

Then

$$v_2 = (900 \text{ km/h}) (0.637)/0.707 = 811 \text{ km/h}.$$

The distance traveled during the second leg is

$$d_2 = (811 \text{ km/h})(3.0 \text{ h}) = 2433 \text{ km}.$$

(a) The average speed is (total distance)/(total time) = (1784 km + 2433 km)/(5.0 h) = $\boxed{843 \text{ km/h}}$.

(b) To find the average velocity, we must first find the displacement which is the sum of the displacements from the two legs of the flight.

$$\vec{r} = \vec{r}_1 + \vec{r}_2 = -d_1 \hat{j} + d_2 \cos(45°)\left(-\hat{i} - \hat{j}\right) = -d_2 \cos(45°)\hat{i} - \left[d_1 + d_2 \cos(45°)\right]\hat{j}$$

$$\vec{r} = -(2433 \text{ m})\cos(45°)\hat{i} - \left[1784 \text{ km} + (2433 \text{ km})\cos(45°)\right]\hat{j} = -(1720 \text{ km})\hat{i} - (3504 \text{ km})\hat{j}$$

The average velocity is

$$\vec{v} = \frac{\vec{r}}{t_{\text{tot}}} = \frac{-(1720 \text{ km})\hat{i} - (3504 \text{ km})\hat{j}}{5 \text{ h}} = \boxed{-\left(344\hat{i} + 701\hat{j}\right) \text{ km/h}}$$

which is equivalent to

$$v = \sqrt{(344)^2 + (701)^2} \text{ km/h} = \boxed{781 \text{ km/h}}$$

$$\theta_{\text{ref}} = \tan^{-1}\left(\frac{-701 \text{ km/h}}{-344 \text{ km/h}}\right) = 64° \text{ S of W} = \boxed{26° \text{ W of S}}$$

(c) From part (b) the final position is

$$\vec{r} = \boxed{-(1720 \text{ km})\hat{i} - (3504 \text{ km})\hat{j}}$$

which is equivalent to

$$r = \sqrt{(1720)^2 + (3504)^2} \text{ km} = \boxed{3900 \text{ km}}$$

$$\phi_{\text{ref}} = \tan^{-1}\left(\frac{-3504 \text{ km/h}}{-1720 \text{ km/h}}\right) = 64° \text{ S of W} = \boxed{26° \text{ W of S}}$$

Answers to Practice Quiz

1. (b) 2. (a) 3. (c) 4. (e) 5. (b) 6. (e) 7. (d) 8. (e) 9. (c) 10. (e)

Answers to Practice Problems

1. 24.4 m
2. 33.6 m/s
3. 2.6 m/s
4. 28.2 m/s^2
5. −21.9°
6. 70.1
7. 1.98 m/s^2
8. 72.7°
9. 2.43 mph
10. 77.0°

CHAPTER 4

NEWTON'S LAWS

Chapter Objectives

After studying this chapter, you should

1. be able to state, and understand the meaning of, Newton's three laws of motion.
2. be able to apply Newton's laws to simple situations in one and two dimensions.
3. be able to draw free-body diagrams.

Chapter Review

This chapter begins the study of **dynamics**, which is the study of the forces, and other principles, that allow us to know why things move the way they do. Newton's three laws of motion are very powerful. These laws are used throughout physics and it is very important that you understand them well. In this chapter, we only introduce Newton's laws and give a few simple examples. More detailed applications of these laws, involving different kinds of forces, will be left to chapter 5.

4–1 Forces and Newton's First Law

A **force** exists on an object when some agent acts to push or pull that object in some way. Some of the most common examples are forces exerted by gravity, springs, strings, people, friction, electric forces, and magnetic forces. Force is a vector quantity. So every force has a direction and a magnitude that represents the strength of the force. When two or more forces act on the same object, the result of all those simultaneous forces, call the **net force**, is determined by the vector sum of all the forces acting

$$\vec{F}_{\text{net}} = \sum_i \vec{F}_i$$

Newton's first law of motion, also called the law of inertia, pertains to the special case when the net force on an object is zero:

> *When there is no net force acting on an object, that object maintains its motion with a constant velocity.*

In other words, objects naturally move at constant velocity, that is, at constant speed in a straight line. The only way to divert an object from its constant velocity motion is to apply a nonzero net force on the object. This natural tendency to move at constant velocity, that all material objects possess, is called **inertia**. Note also, that this law applies to any constant velocity, including zero (at rest). Therefore, an object at rest will remain at rest unless acted upon by a nonzero net force.

As you learned in the previous chapter, whether an object is at rest depends on the frame of reference in which the velocity is measured. For any object moving with constant velocity, there is always a reference frame in which its velocity is zero; and infinitely many other reference frames moving at constant velocities with respect to it. These constant relative velocity reference frames are called **inertial frames**. It is in such reference frames as these that the law of inertia is observed to hold.

4–2 Newton's Second Law of Motion

The law of inertia says that an object's velocity will change (i.e., it will accelerate) if a nonzero force is applied to it; therefore, **force causes acceleration**. Newton's second law picks up on this theme and tells us, quantitatively, how the acceleration relates to the net force.

> *An object acted upon by a net force accelerates. The acceleration equals the ratio of the net force on the object to its mass.*

As an equation, this law is most commonly written as

$$\vec{F}_{net} = m\vec{a}$$

where m is the mass of the object in question. As the equation suggests, the SI unit of force is the product of the SI units of mass and acceleration; it is called a newton (N): $1\ N = 1\ kg\cdot m/s^2$.

The mass in the equation for Newton's second law is sometimes called the *inertial mass*. Why? Considering just the magnitudes, we can rewrite this equation as $a = F_{net}/m$. This shows that the acceleration is inversely proportional to the mass. Therefore, for a given net force, an object with large mass will experience a small acceleration and vice versa. So, it is the mass that determines how much an object resists acceleration – a lot of mass means a lot of resistance. Resistance to acceleration is just another phrase for tendency to maintain a constant velocity. Thus, **mass is a measure of inertia**; hence, the name "inertial mass."

Example 4–1 Skydiving equipment, including the parachute, typically has a mass of about 9.1 kg. At one point during a jump, an 80.0-kg skydiver is accelerating downward at 1.50 m/s². Determine the net force on the skydiver.

Setting it up:
It is helpful to draw a sketch of the skydiver. The sketch shows the skydiver in the air and the two main forces involved. As indicated, upward is chosen as the positive direction. The following information is given in the problem:

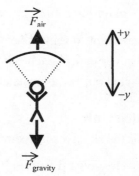

Given: $m_{diver} = 80.0$ kg, $m_{chute} = 9.1$ kg, $\vec{a} = -1.50\ m/s^2\,\hat{j}$
Find: $\vec{F}_{net}$

Strategy: The net force is given by Newton's second law. While it will equal the vector sum of the forces shown, we do not need to take that sum directly.

Working it out:

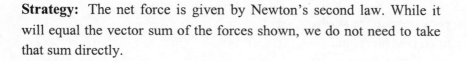

Newton's second law tells us that $\vec{F}_{net} = M\vec{a}$, where M is the total mass of the skydiver and equipment. Since mass is a scalar, we find the total mass by simple addition. Thus,

$$\vec{F}_{net} = \left(m_{diver} + m_{chute}\right)\vec{a} = \left(89.1 \text{ kg}\right)\left(-1.50 \text{ m/s}^2\,\hat{j}\right) = -134 \text{ N}\,\hat{j}$$

What do you think? Suppose the force due to gravity stays the same, while the force due to the air is increased. Would the downward acceleration be greater than, less than, or the same as the given value?

Practice Quiz

1. If the same force is applied to two different objects, the object with greater inertia will have
    (a) greater acceleration   **(b)** smaller acceleration   **(c)** the same acceleration as the other object
    (d) zero acceleration   **(e)** none of the above

2. If two different objects have the same acceleration, the object with greater inertia has
    (a) greater force applied to it
    (b) less force applied to it
    (c) the same force applied to it as the other object has applied to it
    (d) zero force applied to it
    (e) none of the above

3. An object is observed to have an acceleration of 8.3 m/s^2 when a net force of 12.2 N is applied to it. What is its mass?
    (a) 100 kg   **(b)** 0.68 kg   **(c)** 1.5 kg   **(d)** 21 kg   **(e)** 3.9 kg

4–3 Newton's Third Law of Motion

Newton's third law of motion is also known as the law of action and reaction. The words *action* and *reaction* refer to forces and so *force* is the word we'll use:

> *For every force that an agent applies to an object, there is a reaction force equal in magnitude and opposite in direction applied by the object to the original agent.*

It is important to remember that a force and its reaction *always act on different objects*; therefore, these forces *never* cancel each other. Basically, this law says that a single object cannot act on others without being acted on; two objects always interact, applying equal and opposite forces to each other.

Practice Quiz

4. For the skydiver in Example 4–1, what is the reaction force to $\vec{F}_{gravity}$?

    (a) The upward force of the parachute on the skydiver.
    (b) The force that the air exerts on the parachute.
    (c) The gravitational force of the earth on the parachute.
    (d) The gravitational force of the skydiver on the earth.
    (e) The force the parachute exerts on the air.

*4–4 Noninertial Frames

Newton's laws as discussed in this chapter apply in inertial frames. Recall that inertial frames are those with constant relative velocities. In a noninertial frame of reference, such as the frame of an accelerating or rotating observer (detector), Newton's laws are violated because the observer would find accelerating objects on which there is no force; or, alternatively, the observer would infer the existence of forces for which there are no agents. Such a force is called a **fictitious force**. For these reasons, it is usually simplest to analyze physical situations from the point of view of an inertial frame; this is what we will do in the vast majority of cases.

4–5 — 4–6 Using Newton's Laws

Free-body Diagrams

When making a direct application of Newton's second law, certain techniques, or steps, are particularly helpful. The first step is to identify the relevant objects in your problem and all of the forces acting on each object. Once all of the forces are identified, a **free-body diagram** should be drawn for each relevant object.

In a free-body diagram, we isolate the pertinent body, making it a "free body," then we draw all the force vectors acting on this body. In most (but not all) cases of nonrotational motion, it is desirable to idealize the object as a point in your diagram. It is also common to draw the forces with the tails of the vectors on the body (or the idealized point) and the heads pointing away from the body in the associated directions.

Example 4–2 Three boxes are in contact with each other on a frictionless factory table. The boxes are being pushed forward by a mechanical arm as shown (the mechanism controlling the arm is not shown). Draw a free-body diagram of the middle box.

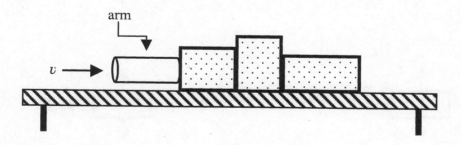

Setting it up/Strategy: We need to isolate the middle box, identify all of the forces acting on it and their directions, then draw the free-body diagram.

Working it out: Let us identify the forces on the middle box. The boxes will be identified by number from left to right. So, box 1 is the left box, box 2 is the middle box, and box 3 is the right box.

(a) Box 1 pushes it forward (to the right), call it the force of box 1 on box 2, $\vec{F}_{12}$.

(b) Because box 2 pushes box 3 forward, we know, by Newton's third law, that box 3 must also push box 2 backward (to the left), call it $\vec{F}_{32}$.

(c) There will be a downward gravitational force on it as well, $\vec{F}_{G2}$.

(d) The table holds it up against gravity's attempt to pull it down. The force that the table exerts on the box is perpendicular to the box and is called the **normal force**, $\vec{F}_{N2}$.

Thus, the free-body diagram of box 2 is

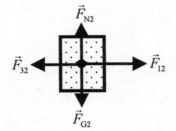

What do you think? Ultimately, the mechanical arm is what causes the boxes, including the middle box, to move. Why isn't the force from the arm in the diagram? On what free-body diagram should the force due to the arm appear?

Practice Quiz

5. For the situation described in Example 4–2, which of the following is a correct free-body diagram for box 3?

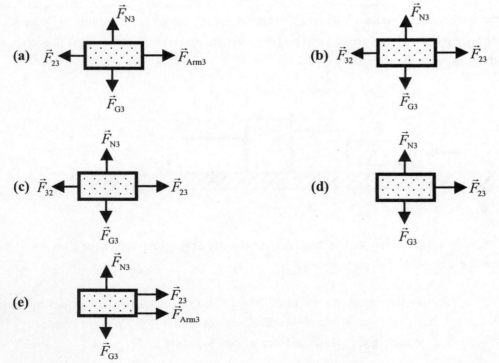

Finding the Motion

Once the free-body diagram has been drawn, you are then in a good position to write out an equation for the net force on the object to go on the left-hand side of $\vec{F}_{net} = m\vec{a}$. Generally, if the problem is two or three dimensional, it is better to write out the net force for each direction using the components of the forces involved. So, for the case in Example 4–2, we could write

$$F_{net, x} = F_{12} - F_{32} = m_2 a_x$$
$$F_{net, y} = F_{N2} - F_{G2} = m_2 a_y = 0$$

The second equation was set equal to zero because the boxes only move horizontally and therefore have no vertical acceleration.

There are two special cases for which we have already studied the resulting motion. The first case is for zero net force. When the net force is zero, the acceleration is also zero; therefore, the motion is constant velocity motion. The second case is that of a nonzero, constant net force. For this latter case, we get motion with constant acceleration. Time dependent net forces complicate matters considerably; only a few such cases will be considered in this text.

Example 4–3 With a full tank of gas, the Pontiac Grand AM SE has a mass of about 1500 kg. Upon pressing the gas pedal, and with good traction, it experiences an average net force of 6.0×10^3 N in the forward direction. If it starts from rest, how fast will the car be moving 6.7 seconds later?

Setting it up: A free-body diagram of the car is shown on the right; $\vec{F}_G$ is the force due to gravity, $\vec{F}_N$ is the normal force from the road, and $\vec{F}$ is the force propelling the car forward (from friction). Air resistance is neglected.

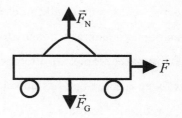

The following information is given in the problem:
Given: $m = 1500$ kg, $\vec{F}_{net} = (6000 \text{ N}) \hat{i}$, $v_0 = 0$, $t = 6.7$ s ; **Find**: v

Strategy: We seek the speed of the car at a particular time. Noting that the net force is constant, we then also know that the acceleration is constant. This means that we can use the equations for constant acceleration that we have used in previous chapters to describe this motion.

Working it out: In Chapter 2, we introduced a system of four equations to describe one-dimensional motion with constant acceleration. From among those expressions, the one involving both the final speed and time (that would allow us the calculate the speed at a given time) is $v = at + v_0$. We are clearly given the initial speed; therefore, what remains is to determine the acceleration.

Newton's second law involves the acceleration, so we turn to it for help. According to this law

$$F_{net} = ma \quad \Rightarrow \quad a = F_{net} / m$$

Since we are given the net force and the mass, the acceleration is effectively determined. Putting this result into the equation for the speed, and using the fact that $v_0 = 0$, gives

$$v = \frac{F_{net}}{m}t = \frac{6000 \text{ N}}{1500 \text{ kg}}(6.7 \text{ s}) = 27 \text{ m/s}$$

What do you think? Suppose this car is going to keep going for a while and that the average net force remains the same. However, you now have to consider the fact that the automobile will be using up fuel. How does this consideration complicate matters? Does the acceleration remain constant even though the net force does?

Practice Quiz

6. Under the action of a constant net force, a 3.1-kg object moves a distance of 15 m in 8.99 s starting from rest. What is the magnitude of this net force?

(a) 0.37 N   (b) 1.2 N   (c) 420 N   (d) 5.2 N   (e) 0.58 N

Reference Tools and Resources

I. Key Terms and Phrases

dynamics the branch of physics that studies force and the causes of various types of motion

net force the vector sum of all forces acting on an object

inertia an object's natural tendency to move with constant velocity

inertial frames the frames of reference in which the law of inertia holds

fictitious force a force that has no source whose existence is inferred due to observations from a noninertial reference frame

free-body diagram a diagram of an isolated object showing all the force vectors acting on the object

normal force the component of the contact force on a surface that is perpendicular to the surface

II. Important Equations

Name/Topic	Equation	Explanation
Net force	$\vec{F}_{net} = \sum_i \vec{F}_i$	The net force on an object is the vector sum of all forces acting on it.
Newton's second law	$\vec{F}_{net} = m\vec{a}$	Newton's second law relates the net force on an object to the acceleration that results from that force.

III. Know Your Units

Quantity	Dimension	SI Unit
Force (F)	$[MLT^{-2}]$	N

Practice Problems

1. A 6.28-kg box is released on an incline and accelerates down the incline at 1.72 m/s². To the nearest tenth of a degree, what is the angle of the incline?

2. A 0.8-kg mass is hung from a string in an airplane awaiting take off. As the plane accelerates, the string makes an angle of 9 degrees with the vertical. To the nearest hundredth of a m/s², what is the acceleration of the plane?

3. A fisherman yanks a fish straight up out of the water with an acceleration of 1.87 m/s² using very light fishing line that has a "test" value of 6 pounds. The fisherman, unfortunately, loses the fish as the line snaps. To the nearest hundredth of a kilogram what is the minimum mass of the fish?

4. A 0.145-kg baseball traveling 33 m/s strikes the catcher's mitt, which, in bringing the ball to rest, recoils backward 21.8 cm. To the nearest tenth of a newton, what was the magnitude of the average force applied by the ball on the glove?

5. The box in the figure has a mass of 70 kg. To the nearest newton, what is the tension in the rope?

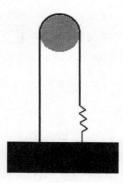

6. If the spring in the figure in Problem 5 has a spring constant of 465 N/m, how far is it stretched, to the nearest tenth of a cm?

7. An elevator (mass 1856 kg) is to be designed so that the maximum acceleration is 2 m/s². To the nearest newton, what is the maximum force the motor needs to exert on the supporting cable?

8. A 1478-kg car pulls a 104-kg trailer via a chain. The combination has an acceleration of 1 m/s². To the nearest newton, what is the tension in the chain?

9. A 13,077-kg helicopter accelerates upward at 0.37 m/s² while lifting a 2313-lb car. To the nearest newton, what is the lift force exerted by the air on the rotors?

10. Three ropes are pulling on a ring. $F_1 = 161$ newtons and $F_2 = 58$ newtons. To the nearest tenth of a newton, what must F_3 equal if the ring is not to move?

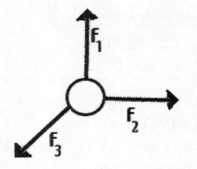

Selected Solutions to End of Chapter Problems

17. Let the additional force be $\vec{F}_3$. Then $F_x = F_{x1} + F_{x2} + F_{x3} = ma_x$ and $F_y = F_{y1} + F_{y2} + F_{y3} = ma_y$.

Because $a_x = 0$ we have that

$$F_{x1} + F_{x2} + F_{x3} = 0 \Rightarrow F_{x3} = -F_{x1} - F_{x2} = -0.50 \text{ N} - (2.0 \text{ N}) \cos 135 = 0.91 \text{ N}$$

For the y component, we have

$$F_{y3} = ma_y - F_{y1} - F_{y2} = (2.5 \text{ kg})(1.5 \text{ m/s}^2) - 0 - (2.0 \text{ N}) \sin 135 = 2.3 \text{ N}.$$

Therefore, the additional force is

$$\vec{F}_3 = \boxed{(0.91 \text{ N})\hat{i} + (2.3 \text{ N})\hat{j}}.$$

21. We need to look at horizontal forces only. The tension in the pulled rope must be equal to the force the father exerts, therefore, $\boxed{T = F}$. If we take both sleds as the object, we get the upper force diagram shown. Then for horizontal motion, we have

$$F = T \cos 30° = (m + m)a = 2ma,$$

which gives,

$$a = \frac{T \cos(30°)}{2m}$$

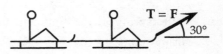

If we take the second sled as the object we get the lower force diagram shown. Then,

$$T_2 = ma = \tfrac{1}{2}T \cos(30°) = \boxed{(0.433)F}.$$

27. (a) The space shuttle's gravitational force on the Earth;

  (b) the skates exert a frictional force and a normal force on the ice and the skater exerts a gravitational force on the Earth;

  (c) none

35. Professor B will be able to tell that he is accelerating because he will feel the force exerted by the back of the seat. (At very small accelerations, this may not be evident.) If Professor B simply looks at the change in position of Professor A, he would think that Professor A is accelerating. This apparent acceleration would be $\boxed{0.70 \text{ m/s}^2}$ in the $-x$-direction. This is not a real acceleration, because there is no horizontal force on Professor A.

55. (a) There is a gravitational force on the person: $F_g = \boxed{mg \text{ down.}}$

There is also a normal force from the elevator floor. Because the acceleration is zero, the normal force must be equal and opposite to the weight: $F_N = F_g = \boxed{mg \text{ up}}$.

(b) We choose up as the positive direction. From $\Sigma \vec{F} = m\vec{a}$ we can write $F_N - F_g = mg$, which gives $F_N = F_g + mg = 2mg$. So, the forces are:

Gravity: $\boxed{F_g = mg \text{ down}}$

Normal Force: $\boxed{F_N = 2mg \text{ up}}$.

(c) Since the acceleration is now g down, $F_N = 0$ and the only force is Gravity: $\boxed{F_g = mg \text{ down}}$.

71. (a) We do a dimensional analysis of $F_d = Av^2$: $[F_d] = [A][v^2]$.

The dimension of force is $[MLT^{-2}]$, and for speed we have $[LT^{-1}]$. So, $[MLT^{-2}] = [A][LT^{-1}]^2$, which gives

$[A] = \boxed{[ML^{-1}] \text{ with units of kg/m}}$.

(b) dv/dt is the acceleration of the parachutist. So, by Newton's second law

$dv/dt = F/m = (Av^2 - mg)/m = \boxed{Av^2/m - g}$.

(c) Terminal velocity is reached when the acceleration equals zero. Using the result of part (b), we have,

$$dv/dt = 0 \quad \Rightarrow \quad (A/m)v_t^2 - g = 0$$

which gives

$$v_t = \sqrt{mg/A} \, .$$

Answers to Practice Quiz
1. (b) **2.** (a) **3.** (c) **4.** (d) **5.** (d) **6.** (b)

Answers to Practice Problems
1.	10.1°	**2.**	1.55 m/s^2
3.	2.29 kg	**4.**	362.2 N
5.	343 N	**6.**	73.8 cm
7.	21,901 N	**8.**	104 N
9.	143,675 N	**10.**	171.1 N

CHAPTER 5

APPLICATIONS OF NEWTON'S LAWS

Chapter Objectives

After studying this chapter, you should

1. be able to apply Newton's second law to situations involving weight, normal force, friction, tension, and drag.
2. be able to apply Newton's second law when motion takes place on an inclined plane.
3. be able to apply Newton's second law when an object moves in circular motion.

Chapter Review

This chapter continues the study of Newton's laws by investigating ways in which these laws are used in common applications. To accomplish this goal, several commonly encountered forces are either introduced or further explained in this chapter.

5–1—5–2 Common Forces Revisited and Friction

The first of the common forces we mention is gravity. Gravity is responsible for an object's **weight**. In Chapter 4, we have already noted that near Earth's surface the effect of gravity is to cause objects to fall with a constant acceleration $\vec{g}$. This effect occurs because near Earth's surface, the force of gravity is constant. In light of Newton's second law, this force, the object's weight, can be written as

$$\vec{W} = m\vec{g}.$$

Therefore, the direction of an object's weight is always vertically downward (toward Earth's center).

Another common force is the **tension** in a rope (or cord of some sort). For ropes that are of negligible mass, the force of tension $\vec{T}$ is transmitted throughout the rope undiminished. Tensions always pull on objects, never push. When two surfaces are in contact, the force that one surface exerts on the other can often be resolved into components parallel to the surface and perpendicular to the surface. The perpendicular component is the normal force $\vec{F}_N$ ("normal" means perpendicular). The normal force on an object always pushes, never pulls. Effectively, the normal force is a measure of how tightly pressed together the two surfaces are.

The parallel component of the contact force between two surfaces is called **friction**. This force becomes a factor when one surface either slides or attempts to slide across the other. The direction of the force of friction always opposes the motion (or attempted motion) between the surfaces. When the surfaces are sliding across each other we call the force **kinetic friction**. When there is only attempted motion, that is halted by friction, we call it **static friction**.

For kinetic friction, the magnitude of the frictional force, f_k, is directly proportional to the magnitude of the normal force between the surfaces. The measured proportionality factor is called the *coefficient of kinetic friction*, μ_k. Thus, we have

$$f_k = \mu_k F_N$$

Note that this equation relates only the magnitudes of $\vec{f}_k$ and $\vec{F}_N$ because they are not in the same direction; they are perpendicular to each other. For static friction the force takes on a range of values depending on the strength of the force it opposes. Static friction will cancel out any force trying to slide two surfaces across each other up to a maximum magnitude beyond which static friction is overcome. The magnitude of this maximum force of static friction is also directly proportional to the magnitude of the normal force. The measured proportionality factor is called the *coefficient of static friction*, μ_s. Thus, we have

$$0 \le f_s \le \mu_s F_N .$$

Typically, for a given pair of surfaces we have $\mu_k < \mu_s$.

Example 5–1 A 5.32-kg box is held stationary on a ramp inclined at 40.0 degrees to the horizontal by a cord that is attached to a vertical wall. The length of the cord is parallel to the ramp, and it provides just enough force to hold the box in place. If the coefficient of static friction between the box and the ramp is 0.101, what is the tension in the cord?

Setting it up:
The left-hand diagram shows the box on the ramp being held in place by the cord. The right-hand sketch is the free body diagram of the box. The information given in the problem is the following
Given: $m = 5.32$ kg, $\theta = 40.0°$, $\mu_s = 0.101$
Find: T

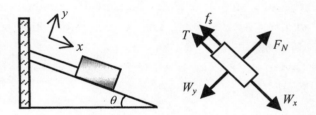

Strategy:
The box has zero acceleration, so we must apply the equilibrium condition, $\sum \vec{F} = 0$, in each direction if necessary. Because the tension is *just* enough to balance the forces, static friction must have its maximum value.

Working it out:
In the diagram we have chosen the x direction to be parallel to the incline and the y direction to be perpendicular to it. Since the tension acts in the x direction, we start by applying the equilibrium condition to the x direction;

$$W_x + T_x + f_{s,\text{max}} = 0 \;\Rightarrow\; W_x - T - \mu_s F_N = 0 \;\Rightarrow\; T = W_x - \mu_s F_N$$

So, to get the desired result for T, we must find the normal force F_N and the x component of the weight.

The x and y components of the weight can be determined by dividing the vertically downward weight vector into components parallel and perpendicular to the incline. As shown in Example 5–8 in your textbook, these components are given by $W_x = mg\sin\theta$ and $W_y = mg\cos\theta$. To determine the normal force, we apply the equilibrium condition to the y direction.

$$F_N + W_y = 0 \quad \Rightarrow \quad F_N - mg\cos\theta = 0 \quad \therefore \quad F_N = mg\cos\theta$$

Now we can return to our equation for the tension and write

$$T = mg\sin\theta - \mu_s mg\cos\theta = mg(\sin\theta - \mu_s\cos\theta).$$

Obtaining the final result gives

$$T = (5.32 \text{ kg})(9.80 \text{ m/s}^2)\left[\sin(40.0°) - (0.101)\cos(40.0°)\right] = 29.5 \text{ N}.$$

What do you think?   At what angle would the tension in the cord be zero? At what angle would the tension have its maximum value? What is that maximum value?

Example 5–2 While rearranging the living room, a student attaches a cord to a 27.5-kg sofa and pulls it across the room accelerating it at 0.150 m/s². If the cord makes an angle of 55.0° above the horizontal, and the coefficient of kinetic friction between the sofa and the floor is 0.240, what is the tension in the cord?

Setting it up:

A free body diagram of the sofa is drawn on the right. The sofa only accelerates horizontally, so there is zero vertical acceleration. The given information is the following.

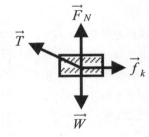

Given: $m = 27.5$ kg, $a = 0.150$ m/s², $\phi = 55.0$, $\mu_k = 0.240$

Find: T

Strategy:

In order to find the desired force, we need to apply Newton's second law. We do so separately in the x and y directions.

Working it out:

Applying Newton's second law in the x direction, with the positive x direction to the right, yields the equation

$$\sum F_x = f_{k,x} + T_x = ma_x \quad \Rightarrow \quad \mu_k F_N - T\cos\phi = -ma \quad \therefore \quad T = \frac{\mu_k F_N + ma}{\cos\phi}$$

The only unknown in this equation is the normal force. We can examine the normal force by applying Newton's second law in the y direction. Since there is no acceleration in the vertical direction, the net force in that direction must equal zero.

$$\sum F_y = F_{N,y} + W_y + T_y = 0 \quad \Rightarrow \quad F_N - mg + T\sin\phi = 0 \quad \therefore \quad F_N = mg - T\sin\phi$$

Substituting this into the equation for T gives

$$T = \frac{\mu_k(mg - T\sin\phi) + ma}{\cos\phi}.$$

Solving this for T then gives

$$T = \frac{m(\mu_k g + a)}{\cos\phi + \mu_k \sin\phi},$$

which produces the final result,

$$T = \frac{(27.5 \text{ kg})\left[(0.240)(9.80 \text{ m/s}^2) + 0.150 \text{ m/s}^2\right]}{\cos 40.0° + (0.240)\sin 40.0°} = 74.8 \text{ N}.$$

What do you think? What would the acceleration be if the student pushed down on the sofa at the same angle rather than pulling up on it?

Practice Quiz

1. A person pushes an object across a room, on a horizontal floor, by applying a horizontal force. If, instead, he pushed it with a force that made an angle θ below the horizontal, the object would have experienced

    (a) greater friction   **(b)** less friction   **(c)** the same frictional force

    (d) zero friction   **(e)** none of the above

2. A person pulls an object across a room, on a horizontal floor, by applying a horizontal force. If, instead, she pulled it with a force that made an angle θ above the horizontal, the object would have experienced

    (a) greater friction   **(b)** less friction   **(c)** the same frictional force

    (d) zero friction   **(e)** none of the above

3. The coefficient of kinetic friction represents

    (a) the force that one surface applies to another when they slide across each other.

    (b) the force that one surface applies to another when they do not slide across each other.

    (c) the force that one surface applies to another that is perpendicular to the interface between them.

    (d) the ratio of the normal force to the force of friction between two sliding surfaces.

    (e) none of the above.

4. Two people pull on opposite ends of a rope. If each person pulls with 30 N, what is the tension in the rope?

    (a) 60 N   **(b)** 30 N   **(c)** 15 N   **(d)** 7.5 N   **(e)** 0 N

5. An object is at rest on a horizontal table. The normal force exerted by the table on the object

    (a) points vertically up   **(b)** points vertically down   **(c)** is parallel to the table

    (d) is zero   **(e)** none of the above

6. An object has a mass of 3.25 kg. Its weight is

    (a) 3.25 kg   (b) 3.25 N   (c) 3.02 N   (d) 7.15 N   (e) none of the above

7. An object is held at rest on a surface inclined at an angle θ with the horizontal. If the force holding it in place is parallel to the surface, the normal force exerted by the surface on the object

    (a) equals its weight   (b) is greater than its weight   (c) is less than its weight

    (d) is zero   (e) none of the above

5–3 Drag Forces

In the previous section, we discussed friction between solid surfaces. When an object moves through a fluid (a liquid or a gas such as air) it experiences a frictional force called **drag**. As with kinetic friction, the direction of the drag force, $\vec{F}_D$, is always opposite to the direction of motion of the object. The magnitude of the drag force depends on a number of quantities including the size and shape of the moving object, the density of the medium through which the object moves, and the speed of the object. For many common situations, at moderate speeds, drag can be modeled by a fairly compact expression

$$F_D = \tfrac{1}{2}\rho A C_D v^2,$$

where ρ is the density of the fluid, A is the cross sectional area of the moving object, and C_D is its **drag coefficient**. The drag coefficient is a dimensionless number that accounts for the object's shape. An object with a small value of C_D is highly aerodynamic and experiences less drag than an otherwise equivalent object (in mass and cross sectional area) with a larger value of C_D.

One of the effects of drag is to halt the downward acceleration of falling objects. As the object's speed increases, due to gravitational acceleration, the drag force also increases. If the fall is not interrupted, there will come a time in which the drag force balances the weight; $F_D = mg$. Once this balance is reached, the object stops accelerating and has reached terminal speed v_t, where

$$v_t = \left(\frac{2mg}{\rho A C_D}\right)^{1/2}.$$

Example 5–3 Two objects of identical size and shape are dropped from the same height in air. Their cross sectional area is 0.030 m², and they have a drag coefficient 0.270. One object has a mass of 0.300 kg and the other of 0.520 kg. What is the acceleration of each object at the (different) times that they are falling at a speed of 1.30 m/s if the density of air is 1.21 kg/m³?

Setting it up:

Since they have identical size and shape, we know that they have the same cross sectional area and drag coefficient. Therefore, the quantity $(1/2)\rho A C_D$ is the same for both objects (remember, ρ is the density of air in this case). The given information is

Given: $A = 0.230$ m², $C_D = 0.370$, $m_1 = 0.300$ kg, $m_2 = 0.520$ kg, $v = 1.30$ m/s, $\rho = 1.21$ kg/m³

Find: a_1 and a_2

Strategy:

The only forces here are gravity pulling down and drag pushing upward. We need to use Newton's second law to get an expression for the acceleration of each object. For convenience, let's take the downward direction as positive.

Working it out:

For each object we can write the drag force as

$$F_D = \tfrac{1}{2}\rho A C_D v^2 = \tfrac{1}{2}\left(1.21\tfrac{kg}{m^3}\right)\left(0.230 \text{ m}^2\right)\left(0.370\right)v^2 = \left(0.0515\tfrac{kg}{m}\right)v^2$$

Applying Newton's second law gives

$$\sum F_y = mg - F_D = ma \quad \therefore \quad a = g - \frac{F_D}{m}.$$

So, the acceleration of particle 1 is

$$a_1 = g - \frac{F_D}{m_1} = 9.80 \text{ m/s}^2 - \frac{\left(0.0515\tfrac{kg}{m}\right)\left(1.30\tfrac{m}{s}\right)^2}{0.300 \text{ kg}} = 9.51 \text{ m/s}^2$$

and for particle 2 it is

$$a_2 = g - \frac{F_D}{m_2} = 9.80 \text{ m/s}^2 - \frac{\left(0.0515\tfrac{kg}{m}\right)\left(1.30\tfrac{m}{s}\right)^2}{0.520 \text{ kg}} = 9.63 \text{ m/s}^2 .$$

Therefore, at a given speed the more massive object has greater acceleration. So, in air, unlike in vacuum, the heavier object will fall faster, all other things being equal.

What do you think? Which body would fall faster if they were of the same mass and drag coefficient, but with different cross sectional area?

Practice Quiz

8. An object falls through the air under the influence of gravity, if its speed doubles, the drag force on it
    (a) doubles.
    (b) is cut in half.
    (c) triples.
    (d) increases by a factor of four.
    (e) decreases by a factor of four.

9. As an object falls through the air, its acceleration
    (a) remains constant.
    (b) decreases.
    (c) increases.
    (d) has the value g.
    (e) [None of the above.]

5–4 Forces and Circular Motion

As discussed in chapter 3, an object in uniform circular motion has a centripetal acceleration given by

$$\vec{a} = -\frac{v^2}{r}\hat{r},$$

where the unit vector $\hat{r}$ points directly away from the center. By Newton's laws, we know that this acceleration must be caused by a force. Therefore, in order for an object to move in uniform circular motion, it must experience a **centripetal force** equal to its mass times its centripetal acceleration,

$$\vec{F} = -\frac{mv^2}{r}\hat{r}.$$

It is important to recognize that centripetal force is not a new kind of force. The above equation merely specifies a condition that must be met by whatever force holds the object in its circular path. In your study of physics, you will find cases in which a centripetal force is caused by gravity, normal force, friction, tension, as well as electrical and magnetic forces.

Above, we described the situation for *uniform* circular motion. If the motion is not at uniform speed, as with the motion of a simple pendulum, there must also be an acceleration that is tangent to the path. Since the velocity itself is always tangent to the path this tangential component of acceleration is

$$a_{tan} = \frac{dv}{dt}.$$

When applying Newton's second law to this situation, separate the problem into radial and tangential components

$$\sum F_{radial} = -\frac{mv^2}{r}, \quad \sum F_{tangential} = m\frac{dv}{dt}.$$

In fact, it is generally the case that the motion along any curved path will have a nonzero net force in each of the radial and tangential directions.

Example 4–4 A car traveling at a speed of 25 km/h comes to a circular section of road. To be safe, the driver takes his foot off the pedal and allows the car's speed to decrease by 1.7 m/s² while rounding the curve. If the radius of the circular section is 12 m, what is the car's acceleration when it first begins to slow down?

Setting it up:

The diagram shows the directions of the car's velocity and the radial and tangential components of its acceleration. Because the car slows down, the tangential component of acceleration must be in the opposite direction of the velocity. The information given is the following.

Given: $v = 25$ km/h, $a_{tan} = -1.7$ m/s², $r = 12$ m

Find: a

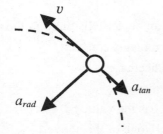

Strategy:

We know that the radial component of the acceleration is given by $a_{rad} = -v^2/r$, and that we are given the tangential component. So, after calculating a_{rad}, we can combine it with a_{tan} to determine the magnitude of the car's acceleration.

Working it out:

Let's first convert the speed to SI units.

$$v = 25 \frac{km}{h} \left(\frac{1000 \text{ m}}{1 \text{ km}} \right) \left(\frac{1 \text{ h}}{3600 \text{ s}} \right) = 6.94 \text{ m/s}.$$

With this, we can determine that

$$a_{rad} = -\frac{(6.94 \text{ m/s})^2}{12 \text{ m}} = -4.02 \text{ m/s}^2.$$

The magnitude of the car's acceleration is found from the components just as with Cartesian components

$$a = \sqrt{a_{rad}^2 + a_{tan}^2} = \left[\left(-4.02 \text{ m/s}^2 \right)^2 + \left(-1.7 \text{ m/s}^2 \right)^2 \right]^{1/2} = 4.4 \text{ m/s}^2.$$

What do you think? Would the magnitude of the acceleration be different if the car sped up at 1.7 m/s² instead of slowed down?

5–5 Fundamental Forces

All of the forces that we have been dealing with such as tension, normal force, and friction result from three basic (or fundamental forces). Gravity is one of these forces and the other two are called the electroweak force and the strong force. The electroweak force represents our current understanding of the relationship between electricity, magnetism, and the weak nuclear force important in some radioactive processes. It is widely believed that, as with the electroweak force, all of the known forces are related, in what some scientists refer to as the *theory of everything*.

Practice Quiz

10. Which of the following is not a fundamental force of nature?

(a) The strong force.

(b) The normal force.

(c) The weak force.

(d) The gravitational force.

(e) The electromagnetic force.

Reference Tools and Resources

I. Key Terms and Phrases

weight the downward force due to gravity

tension the force transmitted through a string or taut wire

friction the parallel component of the contact force between two surfaces

kinetic friction the contact force between two sliding surfaces that opposes their motion

static friction the contact force between two nonsliding surfaces that opposes their attempt to slide

coefficient of kinetic friction the ratio of the force of kinetic friction to the normal force

coefficient of static friction the ratio of the maximum force of static friction to the normal force

drag the frictional force experienced by objects moving through a fluid

drag coefficient a dimensionless quantity accounting for the effect of an object's shape on the drag force

centripetal force any center-pointing force causing an object to move in circular motion

II. Important Equations

Name/Topic	Equation	Explanation
Weight	$\vec{W} = m\vec{g}$	Weight is the downward force due to gravity.
Kinetic friction	$f_k = \mu_k F_N$	The force of friction for sliding objects.
Static friction	$0 \leq f_s \leq \mu_s F_N$	The range of the force of static friction.
Inclined planes	$W_x = mg\sin\theta$ $W_y = -mg\cos\theta$	The components of the weight on an inclined plane that makes an angle θ with the horizontal direction.
Drag force	$F_D = \frac{1}{2}\rho A C_D v^2$	The resistive force on objects moving through a fluid at moderate speed.
Terminal velocity	$v_t = \left(\dfrac{2mg}{\rho A C_D}\right)^{1/2}$	The final constant velocity of objects falling through a fluid at moderate speed.
Circular motion	$\sum F_{radial} = -\dfrac{mv^2}{r}$ $\sum F_{tangential} = m\dfrac{dv}{dt}$	The radial and tangential components on Newton's second law for objects moving along a curved path.

III. Know Your Units

Quantity	Dimension	SI Unit
Coefficient of friction (μ_k)	dimensionless	—
Drag coefficient (C_D)	dimensionless	—

Practice Problems

1. A skier traveling at 36.7 m/s encounters a 21.2 degree slope. If you could ignore friction, how far up the hill does he go, to the nearest meter?

2. If the coefficient of kinetic friction in the previous problem was actually 0.13 and the slope was 30 degrees, how far up the hill does he go, to the nearest meter?

3. You have a mass of 82 kg and are on a 25-degree slope hanging on to a cord with a breaking strength of 159 newtons. To two decimal places, what must the coefficient of static friction be between you and the surface for you to be saved from the fire?

4. In the previous problem if the coefficient of static friction were zero, what would the incline angle have to be in order for the cord not to break, to the nearest tenth of a degree?

5. To the nearest tenth of a degree, what is the banking angle for a 44-m expressway off-ramp curve if it is designed for a speed of 44 km/h?

6. A 1.6-kg ball attached to a 1.8-m cord is swung in a vertical circle. To two decimal places, what is the minimum speed the ball must have at the top of the circle to continue its circular motion?

7. The 53-kg man in the roller coaster car is sitting on a bathroom scale. He is traveling at 37.1 m/s at the point shown and the radius of the vertical coaster track is 54 meters. To the nearest newton, what does the scale read?

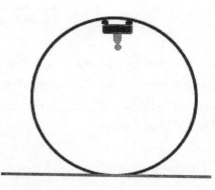

8. What would the answer to the previous problem be if the roller coaster was at the bottom of the track?

9. What would the answer to Problem 7 be if the roller coaster were exactly halfway up the loop and everything else is the same?

10. A 1000-kg car is moving in such a way that its position can be described by the formula:

$$x = 27t^3 + 19t^2 + 2t + 4 \text{ m}$$

What is the accelerating force to the nearest newton at $t = 12$?

Selected Solutions to End of Chapter Problems

13. For the largest value of M, the block is on the verge of slipping down the plane, so the static friction force will be up the plane and maximum, $f_s = f_{s,\max} = \mu_s F_N$. From the force diagram for block M we can write $\Sigma \vec{F} = M \vec{a}$. In terms of components,

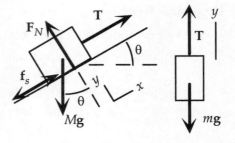

x-component: $T - M_{\max} g \sin\theta + f_{s,\max} = 0$;

y-component: $F_N - M_{\max} g \cos\theta = 0 \Rightarrow F_N = M_{\max} g \cos\theta$.

From the force diagram for block m, Newton's second law gives: $T - mg = 0$; $\Rightarrow T = mg$. Using this result, the x-component equation for block M can be written as

$$mg - M_{\max} g \sin\theta + \mu_s M_{\max} g \cos\theta = 0 \Rightarrow M_{\max}(\sin\theta - \mu_k \cos\theta) = m$$

Solving this we get

$$M_{\max} = \frac{m}{(\sin\theta - \mu_k \cos\theta)} = \frac{3.0 \text{ kg}}{\left[\sin 30° - (0.20)\cos 30°\right]} = \boxed{9.2 \text{ kg}}$$

For the smallest value of M, the block is on the verge of slipping up the plane, so the static friction force will be down the plane and maximum, $f_s = f_{s,\max} = \mu_s F_N$. Applying Newton's second law to the force diagram for block M, the only change from the first case will be in the x-equation, which gives

$$T - M_{\min} g \sin\theta - f_{s,\max} = 0$$

Substituting in known results for T and $f_{s,\max}$ gives

$$mg - M_{\min} g \sin\theta - \mu_s M_{\min} g \cos\theta = 0 \Rightarrow M_{\min}(\sin\theta + \mu_k \cos\theta) = m$$

Solving this we get

$$M_{\min} = \frac{m}{(\sin\theta + \mu_k \cos\theta)} = \frac{3.0 \text{ kg}}{\left[\sin 30° + (0.20)\cos 30°\right]} = \boxed{4.5 \text{ kg}}$$

If $M = 6.0$ kg, the mass is within the range $M_{\min} < M < M_{\max}$, so the block will remain at rest. To determine whether the force of static friction is up or down the plane, we need to determine the critical value of M, M_{crit}, for which $f_s = 0$ and compare it to 6.0 kg. For this case, the x-equation for block M becomes

$$mg - M_{\text{crit}} g \sin\theta + f_s = 0 \Rightarrow mg - M_{\text{crit}} g \sin\theta = 0$$

This gives

$$M_{\text{crit}} = m/\sin\theta = 3.0 \text{ kg}/\sin 30° = 6.0 \text{ kg}$$

Seeing that $M_{crit} = M = 6.0 \text{ kg}$, we can conclude that the answer we seek is $f_s = \boxed{0}$.

27. We write $\Sigma \vec{F} = M\vec{a}$ from the force diagram for the skier, with $\vec{a} = 0$:

x-component: $Mg \sin\theta - f_k = 0$

y-component: $F_N - Mg \cos\theta = 0$

Thus, from the x-component equation: $Mg \sin\theta = f_k = \mu_k F_N$

and from the y-component equation: $F_N = Mg \cos\theta$

This gives,

$$Mg \sin\theta = \mu_k Mg \cos\theta$$

Solving for the coefficient of friction we get

$$\mu_k = (\sin\theta)/(\cos\theta) = \tan\theta = \tan 22° = \boxed{0.40}$$

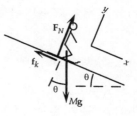

33. Until the box moves, friction is static, opposing the impending motion.

(a) Newton's second law from the force diagram for the box gives:

x-component: $F \cos\theta - f_s = 0$

y-component: $F \sin\theta + F_N - mg = 0$.

When the box is on the verge of moving the force of static friction is maximum. So,

$$f_s = f_{s,max} = \mu_s F_N.$$

The y-component equation gives: $F_N = mg - F \sin\theta$

When this is put into the x-component equation, we get

$$F \cos\theta - \mu_s(mg - F \sin\theta) = 0$$

which we can solve for F to get

$$F = \mu_s mg/(\cos\theta + \mu_s \sin\theta)$$
$$= 0.75(50 \text{ kg})(9.8 \text{ m/s}^2)/(\cos \theta + 0.75 \sin \theta)$$
$$= \boxed{(370 \text{ N}) / (\cos\theta + 0.75 \sin\theta)}.$$

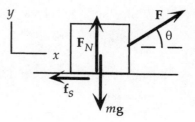

(b) To find the angle at which F is a minimum, we set $dF/d\theta = 0$;

$$dF/d\theta = \mu_s mg (-1)(-\sin \theta + \mu_s \cos\theta)/(\cos\theta + \mu_s \sin\theta)^2 = 0.$$

This is true when

$$\sin \theta_{min} = \mu_s \cos \theta_{min} \quad \Rightarrow \quad \tan \theta_{min} = \mu_s = 0.75$$

which gives $\theta_{min} = \boxed{37°}$.

The force F is then

$$F_{min} = \mu_s mg / [\cos(\tan^{-1} \mu_s) + \mu_s \sin(\tan^{-1} \mu_s)]$$
$$= (370 \text{ N})/[0.80 + 0.75(0.60)] = \boxed{2.9 \times 10^2 \text{ N}}.$$

A minimum value occurs because at small angles the normal force, F_N, and thus the friction force, f_s, are large, which requires a large force F. At angles near 90°, F_N is small, creating a small friction force; however, the small horizontal component of F requires the magnitude of F to be large in this case. So, there is an angle where the decrease in F_N is balanced by the decrease in horizontal component, and thus a minimum value of F.

45. Terminal speed occurs when the drag force equals the weight

$$mg = \tfrac{1}{2}\rho A C_D v^2$$

which shows that

$$v = (\text{constant}) \times m^{1/2}.$$

With $m_1 = 90$ kg, $v_1 = 6.0$ m/s, and $m_2 = 60$ kg, the terminal speed v_2 of the 60-kg person must satisfy

$$v_1/v_2 = (m_1/m_2)^{1/2}$$

which gives

$$v_2 = (m_2/m_1)^{1/2}\, v_1 = (60 \text{ kg}/90 \text{ kg})^{1/2}\,(6.0 \text{ m/s}) = \boxed{5 \text{ m/s}}.$$

55. While the stone does not slide on the turntable, the static friction force provides the centripetal acceleration. Therefore,

$$f_s = ma_r = mR\omega^2.$$

Thus a larger friction force is needed at larger R. At the critical distance, the friction force is maximum

$$f_s = f_{s,\max} = \mu_s mg$$

Where we have set $F_N = mg$ because these are the only vertical forces. So, we have

$$\mu_s mg = mR\omega^2$$

Which gives

$$\mu_s = \frac{R\omega^2}{g} = \frac{(0.21 \text{ m})\left(33 \tfrac{\text{rev}}{\text{min}}\right)^2 \left(2\pi \tfrac{\text{rad}}{\text{rev}}\right)^2 \left(\tfrac{1 \text{ min}}{60 \text{ s}}\right)^2}{9.8 \text{ m/s}^2} = \boxed{0.26}$$

73. For a cable of length L, the radius of the circle is L. The tension in the cable provides the centripetal acceleration. Newton's second law for the radial direction, $\Sigma F_r = ma_r$, gives

$$T = mr\omega^2 = mL\omega^2$$

The angular velocity can be determined to be

$$\omega = \frac{1 \text{ rev}}{1.4 \text{ s}} = \frac{2\pi \text{ rad}}{1.4 \text{ s}} = 4.488 \text{ rad/s}$$

Therefore, we have

$$T = (5.0 \text{ kg})(1.2 \text{ m})(4.488 \text{ rad/s})^2 = \boxed{1.2 \times 10^2 \text{ N}}$$

Answers to Practice Quiz

1. (a) **2.** (b) **3.** (e) **4.** (b) **5.** (a) **6.** (e) **7.** (c) **8.** (d) **9.** (b) **10.** (b)

Answers to Practice Problems

1.	190 m	**2.**	112 m
3.	0.25	**4.**	11.4°
5.	19.1°	**6.**	4.20 m/s
7.	831 N	**8.**	1871 N
9.	1,351 N	**10.**	1,982,000 N

CHAPTER 6

WORK AND KINETIC ENERGY

Chapter Objectives

After studying this chapter, you should

1. be able to calculate the kinetic energy of a particle.
2. be able to determine the work done by constant forces and forces that vary with position.
3. be able to apply the work-energy theorem.
4. be able to distinguish between conservative and nonconservative forces.
5. be able to understand and use the concept of power.

Chapter Review

In the previous chapters, you learned how to analyze mechanical situations using Newton's laws of motion. In this chapter, you begin to learn other important concepts, consistent with Newton's laws, that can make the analysis much easier to perform. This improvement is especially true when the number of applied forces is large, when they vary with distance or time, or when they are not accurately known. Key to this new method of analysis is the concept of **energy**. Energy is a property of physical objects and systems. The study of energy, its different forms, and the processes by which these forms transfer one to another is one of the most important aspects of physics.

6–1 Kinetic Energy, Work, and the Work-Energy Theorem

When the net force on an object acts in such a way as to displace the object, we say that the net force does **work**, W_{net}, on the object. This work can be positive, negative, or zero depending on how the direction of the force relates to the direction of the displacement. If the force is constant and acts along the same line (such as the x direction) as the displacement, the work done by this net force is the product of the net force and the displacement

$$W_{net} = F_{net} \, \Delta x \, .$$

The result of this work is to accelerate the object in such a way that its speed changes. This change in speed can be represented by a change in the object's **kinetic energy** K, where

$$K = \tfrac{1}{2} m v^2$$

and represents the energy that an object has due to its motion. These facts lead to what is called the **work-energy theorem**

$$W_{net} = \Delta K \, .$$

So, the work done by the net force on an object produces a change in the object's kinetic energy.

If the net force points in the opposite direction of the displacement, it does negative work on the object causing a decrease in its kinetic energy. If the net force points in the same direction as the displacement, it does positive work on the object causing an increase in its kinetic energy. If the net force does no work on the object, its kinetic energy remains constant. Since work and energy can equal each other, they must have the same units. The SI unit of work and energy is the **joule** (J), where 1 J = 1 N·m. Another unit of energy that is still in common use is the *erg*, where 1 erg = 1 dyne·cm = 10^{-7} J. Because energy is such an important quantity in so many different contexts, you will come across many different units of energy in your scientific studies.

Recall that the net force on an object is the sum of all the forces acting on it. Similarly, the work done by the net force on an object is the sum of the work done by each individual force acting on it. As you will see, it is often very useful to calculate the work done by individual forces. But, keep in mind that it is the work done by the net force that determines the resulting change in the kinetic energy of the object.

Example 6–1 A 3.5-kg object slides 1.7 m down a ramp that is inclined at 35 degrees to the horizontal. If the coefficient of kinetic friction between the object and the ramp is 0.12, **(a)** how much work is done by gravity, **(b)** how much work is done by kinetic friction, and **(c)** what is the kinetic energy of the object if it started from rest?

Setting it up:

The diagram shows the object on the ramp and the forces acting on it. The information given in the problem is the following

Given: $m = 3.5$ kg, $\Delta x = 1.7$ m, $\alpha = 35°$, $v_0 = 0$, $\mu_k = 0.12$
Find: **(a)** W_g, **(b)** W_f, **(c)** K

Strategy:

The diagram helps to show that the force of gravity has a component along the line of the displacement, which is down the incline. Also, we see that the frictional force is along the same line except in the opposite direction. So, we can calculate the work done by these individual forces using $W = F \Delta x$. Once the work for these forces is known, the kinetic energy can be found using the work-energy theorem.

Working it out:

(a) Because the relevant component of the force of gravity is in the same direction as the displacement, we know that gravity does positive work on the object. So,

$$W_g = (mg\sin\alpha)\Delta x = (3.5 \text{ kg})(9.8 \text{ m/s}^2)(1.7 \text{ m})\sin(35°) = 33 \text{ J}$$

(b) By taking Δx to be positive, we indicate the choice of down the incline as the positive direction. This choice means that the force of friction acts in our negative direction. Therefore, $W_f = (-f_k)\Delta x$. This shows that we need to determine the value of the force of kinetic friction. Since the object does not accelerate in the direction perpendicular to the incline, the two forces along that direction must cancel. This implies that

$$F_N = mg\cos\alpha$$

giving for the force of friction,

$$f_k = \mu_k F_N = \mu_k mg \cos\alpha .$$

The work done by friction is then calculated as

$$W_f = -\mu_k (mg\cos\alpha)d = -(0.12)(3.5 \text{ kg})(9.8 \text{ m/s}^2)\cos(35°)(1.7 \text{ m}) = -5.7 \text{ J}$$

(c) As mentioned in part (b), the forces perpendicular to the incline must cancel each other. Therefore, they do not contribute to the net force on the object. As a result, the net force is determined by the parallel component of gravity and kinetic friction. So, the work done by the net force is the sum of the work done by those two forces

$$W_{net} = W_g + W_f = 33.4 \text{ J} - 5.73 \text{ J} = 28 \text{ J} .$$

By the work-energy theorem, we can say that

$$W_{net} = K - K_0 = K = 28 \text{ J} .$$

What do you think? Would the change in kinetic energy be different if it had not started from rest?

Practice Quiz

1. A person pushes an object across a room, on a horizontal floor, by applying a horizontal force. If, instead, he pushed the object through the same horizontal distance with the same magnitude of force at an angle θ above the horizontal, friction would do
   (a) a greater amount of positive work   (b) a greater magnitude of negative work
   (c) a lesser amount of positive work   (d) a lesser magnitude of negative work   (e) zero work

2. When a net amount of positive work is done on an object, you expect the object to
   (a) speed up   (b) slow down   (c) stop   (d) maintain a constant speed   (e) none of these

3. A 93-g bullet is fired from a gun. It emerges from the muzzle with a kinetic energy of 4200 J, what is the muzzle velocity of the gun?
   (a) 390 m/s   (b) 45 m/s   (c) 300 m/s   (d) 210 m/s   (e) 77 m/s

6–2 Constant Forces in More Than One Dimension

When the (constant) force and the displacement are not along the same line, the work done by the force can be calculated by taking the **scalar product** (or dot product) of the force and displacement vectors,

$$W = \vec{F} \cdot \Delta\vec{r} .$$

The scalar product of two vectors is equivalent to multiplying the magnitude of the first vector by the component of the second vector along the direction of the first, or vice versa. That is,

$$\vec{F} \cdot \Delta\vec{r} = F_\| \Delta r = F \Delta r_\| ,$$

where $F_\parallel$ is the component of $\vec{F}$ parallel to $\Delta \vec{r}$ and $\Delta r_\parallel$ is the component $\Delta \vec{r}$ parallel to $\vec{F}$. If the smallest angle between $\vec{F}$ and $\Delta \vec{r}$ is θ, then these parallel components are given by

$$F_\parallel = F\cos\theta \quad \text{and} \quad \Delta r_\parallel = \Delta r \cos\theta .$$

An alternative way to write the scalar product, in terms of the components of $\vec{F}$ and $\Delta \vec{r}$, is

$$\vec{F} \cdot \Delta \vec{r} = F_x \Delta x + F_y \Delta y + F_z \Delta z .$$

So, the scalar product is the sum of the products of the corresponding components of the two vectors. Putting this all together, we can say that the work done by a constant force is given by

$$W = F\Delta r \cos\theta = F_x \Delta x + F_y \Delta y + F_z \Delta z .$$

Example 6–2 A person pushes a 751-N shopping cart full of groceries down the aisle at the local store. The person applies a force of 112 N at an angle of 40.0 degrees below the horizontal. The cart is pushed the full length of a 15.5-m aisle at constant speed. Determine **(a)** the work done by the shopper, **(b)** the work done by gravity, **(c)** the work done by the normal force of the floor on the cart, and **(d)** the work done by various frictional forces during the cart's motion down the aisle.

Setting it up:

The diagram shows the shopping cart, the force exerted by the shopper $\vec{F}$, the weight $m\vec{g}$, the normal force $\vec{F}_N$, and the displacement $\vec{d}$. The information given is the following:
Given: $mg = 751$ N, $F = 112$ N, $\theta_F = -40.0°$, $d = 15.5$ m,
      $v = $ const.
Find: **(a)** W_F, **(b)** W_g, **(c)** W_N, **(d)** W_f

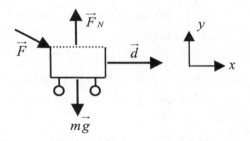

Strategy:

To calculate the work done by the individual forces listed, we need to identify the magnitude of each force and its direction relative to the displacement. We can determine the work done by frictional forces from the total work on the cart.

Working it out:

(a) From the given information, the smallest angle between $\vec{F}$ and $\vec{d}$ is 40.0°. Therefore, the work done by the shopper is

$$W_F = Fd\cos(|\theta_F|) = (112\text{ N})(15.5\text{ m})\cos(40.0°) = 1330\text{ J}$$

(b) From the given situation the angle between the cart's weight and its displacement is 90.0°. So,

$$W_g = mg\cos(90.0°) = 0\text{ J} .$$

This equals zero because cos(90.0°) = 0; or, equivalently, the weight has no component along the line of the displacement.

(c) From the given situation the angle between the normal force and the displacement is 90.0°. So,

$$W_N = F_N \cos(90.0°) = 0 \text{ J}.$$

(d) Because the cart moves with constant speed, the net force on the cart is zero. This fact means that the work done by the net force must also be zero. Thus,

$$W_{net} = W_F + W_g + W_N + W_{fric} = 0.$$

Given that $W_g = W_N = 0$, we can see that

$$W_f = -W_F = -1330 \text{ J}.$$

What do you think?   Could you have determined the work done by the frictional forces in the same way as in parts (a) – (c) instead of using $W_{net} = 0$?

Practice Quiz

4.   A person pushes an object across a room, on a horizontal floor, by applying a horizontal force. If, instead, he pushed it through the same horizontal distance with the same magnitude of force at an angle θ below the horizontal, he would do

    (a) greater amount of positive work   **(b)** lesser amount of positive work   **(c)** zero work

    (d) negative work   **(e)** none of the above

5.   A person pulls a 3.0-kg object across a room, on a horizontal floor, by applying a force of 13 N that makes an angle of 77° above the horizontal. If she does a total of 26 J of work, how far does she pull the object?

    (a) 2.0 m   **(b)** 230 m   **(c)** 3.9 m   **(d)** 10 m   **(e)** 8.9 m

6.   A force, in newtons, given by $\vec{F} = 2.0\hat{i} + 8.0\hat{j}$ acts on a particle that undergoes a displacement $\Delta\vec{r} = 5.0\hat{i} - \hat{j}$ in meters. How much work is done by the force $\vec{F}$?

    (a) 5.1 J   **(b)** 2.0 J   **(c)** 5.0 J   **(d)** –1.0 J   **(e)** 8.2 J

6–3 Forces That Vary with Position

Now that we know how to determine the work done by a constant force, how can we generalize this to varying forces? In particular, we will consider forces that vary with position, which is a common occurrence. First, let's consider the one-dimensional case for which the force and displacement are along the same line of motion (the x direction). The key is to recognize that any force that varies with distance is approximately constant over small enough displacements. For these arbitrarily small displacements $dx\hat{i}$, we can approximate the small amount of work done dW, by the force $\vec{F} = F(x)\hat{i}$ using the expression for a constant force $dW = F dx$. The total work done as the object is displaced from x_0 to x_f is given by the integral over dW,

$$W = \int_{x_0}^{x_f} dW = \int_{x_0}^{x_f} F dx.$$

Which is equivalent to the area under the F vs. x curve. If the force being used is the net force on the object, then we can apply the work-energy theorem

$$\int_{x_0}^{x_f} F_{net}dx = \Delta K .$$

A variable force of special interest is the **Hooke's law** force, $F = -kx$, that exactly describes an ideal spring and approximately describes the behavior of any solid object when it is deformed slightly. In Hooke's law, x is the displacement of the object from the equilibrium position of the system and k characterizes the stiffness of the spring and is called the *spring constant* (or force constant). The work done by the spring on the object as it is displaced from x_0 to x_f is

$$W = \int_{x_0}^{x_f} (-kx)dx = -k \int_{x_0}^{x_f} xdx = -\tfrac{1}{2}kx^2 \Big|_{x_0}^{x_f} = \tfrac{1}{2}kx_0^2 - \tfrac{1}{2}kx_f^2 .$$

Example 6–3 A spring with a spring constant of 2000 N/m is used to cushion objects of moderate weight. If a 75-lb object is placed on one of these springs, how much work does the spring do before it balances the weight of the object?

Setting it up:

Part (a) of the figure shows the object being balanced by the spring. Part (b) is a free-body diagram of the object. The information given is the following.

Given: $k = 2000$ N/m, $mg = 75$ lb

Find: W_s

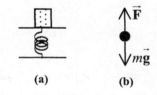

(a)   (b)

Strategy:

Based on the information given here, we need to know how much the spring compresses in order to calculate the work it does while compressing. The spring compresses until it is in equilibrium with the weight of the object.

Working it out:

Let's first convert the weight to newtons.

$$mg = 75 \text{ lb}\left(\frac{4.45 \text{ N}}{1 \text{ lb}}\right) = 333.75 \text{ N} .$$

Now, in equilibrium the net force is zero

$$\sum F = ky - mg = 0 .$$

This fact determines how much the spring compresses

$$ky = mg \implies y = \frac{mg}{k} .$$

Before the object is placed on the spring cushion, the spring is in its equilibrium position; so, $y_0 = 0$. The work done by the spring is then

$$W_s = 0 - \tfrac{1}{2}ky^2 = -\tfrac{1}{2}k\left(\frac{mg}{k}\right)^2 = -\frac{(mg)^2}{2k}.$$

So, we have

$$W_s = -\frac{(333.75\text{ N})^2}{2(2000\text{ N/m})} = -28\text{ J}$$

What do you think? During the object's displacement, does gravity do a greater, lesser, or equal amount of work (in magnitude) as the spring does?

If the force varies in both magnitude and direction, the generalization of the constant force result for arbitrarily small displacements is $dW = \vec{F} \cdot d\vec{r}$. The total work done from position $\vec{r}_A$ to $\vec{r}_B$ is then given by the line integral

$$W = \int_{\vec{r}_A}^{\vec{r}_B} \vec{F} \cdot d\vec{r}.$$

This is the most general case for calculating the work done by a force.

Practice Quiz

7. A spring is held in a compressed position, then released. As the spring uncoils to its equilibrium length it does
   (a) negative work.   (b) positive work.   (c) zero work.   (d) none of the above

8. If it requires an amount of work W to stretch a certain spring by an amount x from its equilibrium length, how much work would it take to stretch it by an amount $2x$?
   (a) $2W$ (b) $\sqrt{2}W$ (c) $\tfrac{1}{2}W$ (d) $\tfrac{1}{4}W$ (e) $4W$

9. A force is given by the function $F = 6.0x^2$, where F is in newtons and x is in meters. How much work is done by this force if it acts on an object while being displaced from $x = 3.0$ m to $x = -3.0$ m?
   (a) -9.0 J   (b) 54 J   (c) 220 J   (d) -110 J   (e) -54 J

6–4 Conservative and Nonconservative Forces

Forces are generally divided into two categories depending on properties of the work that a force does. A force can either be a **conservative force** or a **nonconservative force**. A force is called *conservative* if the work it does on any object is independent of the path that object takes during a displacement. Equivalently, a force is conservative if the work it does around any closed path is zero. Any force that does not meet this condition is a *nonconservative* force.

The primary reason for the two classifications is that when work is done by a conservative force, that work is in some sense stored within the system and can be recovered, usually as kinetic energy. Gravity is a conservative force. When an object is lifted against gravity, a certain amount of negative work is done by gravity on the object. When that object is released, or lowered back down, gravity does an equal amount of positive work. Kinetic friction is a nonconservative force. If you slide a block across a table, friction does a certain amount of negative work on it. When you slide the block back to its original place, friction does even more negative work on it — the work is not recovered in this case. The Hooke's law force that is exerted by a spring is also a conservative force.

6–5 Power

From a practical standpoint, the mere fact that a certain amount of work is done is not always good enough. The question of how long it takes to do this work often determines the practical value of certain devices. The quantity we use to measure how rapidly work is done is called **power** P. Power is the rate at which work is done. We are often interested in the average power

$$P_{av} = \frac{\text{work done}}{\text{time interval}} = \frac{W}{\Delta t} .$$

However, sometimes we require the instantaneous power

$$P = \frac{dW}{dt} .$$

This instantaneous power can also be expressed in terms of the velocity of the object being displaced.

$$P = \frac{dW}{dt} = \frac{\vec{F} \cdot d\vec{r}}{\Delta t} = \vec{F} \cdot \frac{d\vec{r}}{\Delta t} = \vec{F} \cdot \vec{v} .$$

The SI unit of power, derived from the units of energy and time, is called a **watt** (W): 1 W = 1 J/s. Another commonly used unit for power is the *horsepower* (hp): 1 hp = 746 W. When work is being done at a certain average rate P_{av}, the amount of work done in a time interval Δt is $W = P_{av}\Delta t$. A unit of work based on this fact is the kilowatt-hour (kWh) commonly used by electric companies: 1 kWh = 3.6×10^6 J.

Example 6–4 The motor of a chain-linked industrial pulley designed to lift heavy weights is rated to deliver 2.0 kilowatts of power on average. If the mechanism is only 80% efficient, how long will it take to raise a 55-kg crate a height of 12 m at a constant speed?

Setting it up:
We can reduce the problem to the following information:
Given: P_{av} = 2000 W, m = 55 kg, y = 12 m, v = const, eff = 80%
Find: t

Strategy:
The amount of time it takes is related to the average power through the equation $P_{av} = W/t$. Therefore, we will determine the work done and the actual power output.

Working it out:
The time it takes to raise the weight is given by.

$$P_{out} = \frac{W_{pulley}}{t} \quad \Rightarrow \quad t = \frac{W_{pulley}}{P_{out}}$$

Where P_{out} is the actual power output of the device accounting for its level of efficiency. The power output is 80% of power rating

$$P_{out} = 0.8P_{av} = 0.8(2000\,\text{W}) = 1600\,\text{W}\,.$$

The work done by the pulley is given by

$$W_{pulley} = F_{pulley}\,y\,.$$

Because the pulley raises the weight at constant speed (i.e., zero acceleration), we know that the upward force of the pulley must equal the downward weight. Therefore, $F_{pulley} = mg$. So, we can now determine the amount of time to be

$$t = \frac{mgy}{P_{out}} = \frac{(55\,\text{kg})(9.8\,\text{m/s}^2)(12\,\text{m})}{1600\,\text{W}} = 4.0\,\text{s}\,.$$

What do you think? Would the amount of time to lower the weight back to the original location at constant speed be any different?

_____

Practice Quiz

10. If motor A delivers more power than motor B, then

   (a) motor A can do more work than motor B

   (b) motor A takes more time to do the same amount of work as motor B

   (c) motor A takes less time to do the same amount of work as motor B

   (d) both motors take the same amount of time to do the same amount of work

   (e) motor A is more efficient than motor B

11. A lift mechanism delivers a power of 85 W in lifting a 250-N crate at constant speed. What is the speed?

   (a) 0.34 m/s   **(b)** 0.29 m/s   **(c)** 6.0 m/s   **(d)** 9.8 m/s   **(e)** 3.5 m/s

6–6 Kinetic Energy at Very High Speeds

When objects move at speeds that are a significant fraction of the speed of light, $c = 3.00 \times 10^8$ m/s, we must acknowledge that Newton's laws and its consequences produce only low-speed approximations to the more general results from the theory of relativity. According to this theory, the kinetic energy of a particle is given by

$$K = mc^2 \left\{ \left[1 - (v/c)^2 \right]^{-1/2} - 1 \right\}.$$

This result implies that, unless a particle has zero mass, it can never move at the speed of light.

Reference Tools and Resources

I. Key Terms and Phrases

energy a property of physical objects and systems that takes many different forms

kinetic energy the form of energy related to the motion of objects

work-energy theorem the expression $W_{net} = \Delta K$ that relates work and kinetic energy

joule the SI unit of energy

scalar product a form of vector multiplication that produces a scalar quantity measuring the extent to which two vectors are parallel to each other

Hooke's law the force law for an ideal spring

conservative force any force for which the work done is independent of path

nonconservative force any force that is not a conservative force

power the rate at which work is done

watt the SI unit of power

II. Important Equations

Name/Topic	Equation	Explanation
Kinetic energy	$K = \frac{1}{2}mv^2$	The kinetic energy of a particle.
Work-energy theorem	$W_{net} = \Delta K$	The work done by the net force produces a change in kinetic energy.
Work	$W = \vec{F} \cdot \Delta \vec{r}$	The work done by a constant force.
Work	$W = \int_{\vec{r}_A}^{\vec{r}_B} \vec{F} \cdot d\vec{r}$	The work done by a varying force.
Hooke's law	$F = -kx$	Hooke's law is a restoring force directly proportional to a deformation x.
Power	$P = \dfrac{dW}{dt}$	Power is the rate at which work is done.

III. Know Your Units

Quantity	Dimension	SI Unit
Work (W)	$\left[ML^2T^{-2}\right]$	J
Kinetic energy (K)	$\left[ML^2T^{-2}\right]$	J
Power (P)	$\left[ML^2T^{-3}\right]$	W

Practice Problems

1. A man pulls a 10-kg box across a smooth floor with a force of 101 newtons at an angle of 22 degrees and for a distance of 114 meters. To the nearest joule, how much work does he do?

2. Assume the floor in the previous question is angled upward at 8.6 degrees and the man pulls the box up the floor at constant speed. What is the work he does, to the nearest joule?

3. In Problem 2 what is the work done by gravity, to the nearest joule?

4. In Problem 1, if the box starts from rest, what is its final kinetic energy, to the nearest joule?

5. The force on an object is given by

$$F = 2x - 0.28x^2$$

   To the nearest tenth of a joule, what is the work done by this force in moving the object in a straight line from $x = 0.63$ to $x = 2.82$?

6. You are traveling in your 2373-kg car at 13 m/s and wish to accelerate to 21.3 m/s in 4 seconds. To the nearest joule, how much work is required?

7. In the previous problem, what is the average power, to the nearest watt?

8. What is the equivalent (nearest) horsepower in Problem 7?

9. Assume you have a bow that behaves like a spring with a spring constant of 61 N/m. You pull it to a draw of 82 cm. To the nearest joule how much work do you perform?

10. In the previous problem, what is the speed of the 55-gram arrow when it is released, to the nearest tenth of a m/s?

Selected Solutions to End of Chapter Problems

11. We convert the speed units:

(96.6 mi/h)(1.61 × 10³ m/mi)/(3600 s/h) = 43.2 m/s.

(95.3 mi/h)(1.61 × 10³ m/mi)/(3600 s/h) = 42.6 m/s;

The work done by friction is the net work. The work-energy theorem then says,

$$W_f = W_{net} = \Delta K = \tfrac{1}{2} m(v^2 - v_0^2) = \tfrac{1}{2}(0.145 \text{ kg})[(42.6 \text{ m/s})^2 - (43.2 \text{ m/s})^2] = \boxed{-3.73 \text{ J}}.$$

37. Because the force is constant, the work is $W = \vec{F} \cdot \Delta \vec{r}$. The displacement is given by

$$\Delta \vec{r} = \vec{r}_2 - \vec{r}_1 = \left(5m\hat{i} - 4m\hat{j} + 5m\hat{k}\right) - \left(7m\hat{i} - 8m\hat{j} + 2m\hat{k}\right) = -2m\hat{i} + 4m\hat{j} + 3m\hat{k}$$

The work done is therefore

$$W = \left(2N\hat{i} - 5N\hat{j}\right) \cdot \left(-2m\hat{i} + 4m\hat{j} + 3m\hat{k}\right) = -4\text{ J} - 20\text{ J} + 0\text{ J} = \boxed{-24 \text{ J}}$$

45. The work-energy theorem tells us that the work of the spring equals the change in the kinetic energy of the pellet. The work done by the spring is given by $\tfrac{1}{2}kx^2$. So

$$\tfrac{1}{2}kx^2 = \tfrac{1}{2}mv^2 \quad \Rightarrow \quad v = \sqrt{kx^2/m}$$

$$\therefore \quad v = \left[\frac{(60 \text{ N/m})(0.07 \text{ m})^2}{0.004 \text{ kg}}\right]^{1/2} = \boxed{8.6 \text{ m/s}}$$

53. For a constant force, the work done is given by $W_{0\to1} = F\,\Delta x = F(x_f - x_0)$. So,

$$W_{0\to1} = (10 \text{ N})\left[\left(11\text{ m} - 2\tfrac{m}{s} + 0.5\tfrac{m}{s^2}\right) - (11\text{ m})\right] = \boxed{-15 \text{ J}}$$

$$W_{1\to2} = (10\text{ N})\left\{\left[11\text{ m} - 2\tfrac{m}{s}(2\text{s}) + 0.5\tfrac{m}{s^2}(2\text{s})^2\right] - \left[11\text{m} - 2\tfrac{m}{s} + 0.5\tfrac{m}{s^2}\right]\right\} = (10\text{ N})[9\text{ m} - 9.5\text{m}] = \boxed{-5 \text{ J}}$$

The work done by a constant force is independent of the path and thus $\boxed{\text{conservative}}$.

69. Because the velocity is constant, the acceleration is zero. So, Newton's second law gives

$$x\text{-component: } F - \mu_k F_N = 0$$

$$y\text{-component: } F_N - mg = 0$$

The y-component equation gives $F_N = mg$. Using this result in the x-component equation gives

$$F = \mu_k mg = (0.03)(5000 \text{ N}) = 150 \text{ N}.$$

Since the force does not vary, the maximum power will produce the maximum speed

$$P_{max} = Fv_{max} \quad \Rightarrow \quad v_{max} = P_{max}/F$$

$$\therefore v_{max} = (1 \text{ hp})(746 \text{ W/hp})/(150 \text{ N}) = \boxed{5.0 \text{ m/s}}.$$

On an incline, Newton's second law gives

$$x\text{-component: } F - mg\sin\theta - \mu_k F_N = 0;$$

$$y\text{-component: } F_N - mg\cos\theta = 0$$

The y-component equation gives $F_N = mg \cos\theta$. So, the x-component equation becomes

$$F = mg (\sin\theta + \mu_k \cos\theta) = (5000 \text{ N})(\sin 5° + 0.03 \cos 5°) = 585 \text{ N}.$$

As with the previous case, we can now determine the maximum speed from

$$v_{max} = P_{max} / F \quad \Rightarrow \quad v_{max} = (746 \text{ W}) / (585 \text{ N}) = \boxed{1.3 \text{ m/s}}.$$

85. (a) As the block moves up the ramp the three forces on it are

1. Gravity: $mg = (2.6 \text{ kg})(9.8 \text{ m/s}^2) = \boxed{26 \text{ N down}}$.

2. Normal Force: $F_N = mg \cos\theta = (26 \text{ N}) \cos 32° = \boxed{22 \text{ N perpendicular to plane (up)}}$.

    [F_N is equal and opposite to the perpendicular component of the gravitational force]

3. Kinetic Friction: $f_k = \mu_k F_N = (0.25)(22 \text{ N}) = \boxed{5.4 \text{ N parallel to plane (down)}}$.

(b) Each force is constant so $W = F_\parallel \Delta x$.

For gravity, the parallel component is $F_{g,\parallel} = mg \sin\theta$ and is opposite to the displacement. So,

$$W_g = - mg \sin\theta \Delta x = - (26 \text{ N}) \sin 32° (1.3 \text{ m}) = \boxed{-18 \text{ J}}.$$

Because the normal force is perpendicular to the displacement, we can immediately say $W_N = \boxed{0 \text{ J}}$.

As with gravity, the force of friction is opposite to the displacement. So,

$$W_f = - f_k \Delta x = (5.4 \text{ N})(1.3 \text{ m}) = \boxed{-7.0 \text{ J}}.$$

(c) From the work-energy theorem, $W_{net} = \Delta K$, we get

$$W_g + W_N + W_f = \Delta K = K_f - K_0 = -\tfrac{1}{2} mv_0^2$$

$$\therefore \quad -18 \text{ J} - 7.0 \text{ J} = -\tfrac{1}{2}(2.6 \text{ kg})v_0^2 \quad \Rightarrow \quad v_0 = \boxed{4.4 \text{ m/s}}.$$

Answers to Practice Quiz

1. (d) 2. (a) 3. (c) 4. (b) 5. (e) 6. (b) 7. (b) 8. (e) 9. (d) 10. (c) 11. (a)

Answers to Practice Problems

1.	10,676 J	2.	1,671 J
3.	−1,671 J	4.	10,676 J
5.	5.5 J	6.	337,785 J
7.	84,446 W	8.	113 hp
9.	21 J	10.	27.3 m/s

CHAPTER 7

POTENTIAL ENERGY AND CONSERVATION OF ENERGY

Chapter Objectives

After studying this chapter, you should

1. be able to calculate changes in the potential energy functions for gravity and Hooke's law.
2. be able to analyze mechanical problems using the principle of the conservation of total mechanical energy.
3. be able to understand the information contained in energy diagrams.
4. be able to apply the concepts of work and energy in the presence of nonconservative forces.

Chapter Review

In this chapter we continue the study of work and energy and encounter another form of energy called potential energy. Most importantly, in this chapter we introduce the concept of the conservation of energy, which is one of the most important concepts in science.

7–1 Potential Energy and Conservative Forces

In Chapter 6 you were introduced to the idea that when work is done, it goes into changing the kinetic energy of the object, or system, on which the work is done. In that case it was the *net* work done by all forces acting. A similar concept can be applied to individual forces as well. It turns out that when work is done by a conservative force, this work goes into changing another aspect of the system different from its kinetic energy. This different form of energy is called the **potential energy**, U, of the system. This result is a necessary consequence of the fact that the work done by a conservative force only depends on the endpoints of the displacement (that is, it is independent of the path)

$$W_{AB} = U_A - U_B = -\Delta U .$$

So, the change in the potential energy of a system equals the negative of the work done by the conservative force associated with it.

Previously, we developed a conceptual understanding of kinetic energy as the energy "stored in" the motion of objects. What is the analogous descriptive understanding of potential energy? Because the definition of potential energy has to do with the locations of objects in a system, specifically, the work done as an object is moved from one location to another, then we can see that the information represented by the potential energy of the system is "stored in" the knowledge of what these locations are. That is, potential energy is energy associated with the configuration of a system.

For variable forces, considering only the one dimensional case, the relationship between the force and the potential energy function can be seen from the fact that for small displacements we can write $dW = F(x)dx$. Given that we also know that $dW = -dU$, we have by direction comparison

$$F(x) = -\frac{dU}{dx}.$$

One important example of a conservative force is gravity. Near Earth's surface, the work done by gravity as an object moves from height y_0 to height y_f is

$$W_g = mg(y_0 - y_f).$$

This results means

$$-\Delta U = U_0 - U_f = mgy_0 - mgy_f.$$

Therefore, the potential energy for the gravitational force can be written as

$$U(y) = mgy + U_0,$$

where y is the height above a chosen reference level at which $y = 0$ and $U = U_0$. It is important to note that the chosen reference level and its associated value of potential energy are arbitrary. You are free to make those choices according to the convenience of your problem. **Only changes in potential energy are physically significant**, not individual values of U.

A second important example is Hooke's law, the conservative force exerted by an ideal spring. We saw in Chapter 6 that the work done by a spring while being deformed from x_0 to x_f is

$$W = \tfrac{1}{2}kx_0^2 - \tfrac{1}{2}kx_f^2.$$

Therefore, the change in potential energy is given by $-W$,

$$U_f - U_0 = \tfrac{1}{2}kx_f^2 - \tfrac{1}{2}kx_0^2.$$

While, as with the potential energy for gravity, the reference configuration and value are arbitrary, in the case of Hooke's law it is customary to take the undeformed configuration as the one for which $U(x) = 0$. Therefore, we can write the potential energy for Hooke's law as

$$U(x) = \tfrac{1}{2}kx^2$$

with the understanding that the value obtained is actually the *change in* potential energy with respect to the undeformed configuration.

With the concept of potential energy in hand, the work-energy theorem, $W_{net} = \Delta K$, can be written in a new way. For systems in which work is done by conservative forces only, we can write that work as a change in potential energy, $-\Delta U = \Delta K$. This says that any change in kinetic energy is accounted for by an opposite change in potential energy. Expanding this further,

$$U_0 - U_f = K_f - K_0 \implies K_0 + U_0 = K_f + U_f.$$

If we define the quantity $E = K + U$ as the **total mechanical energy**, then we have

$$E_0 = E_f \implies E = \text{constant}.$$

This equation expresses the principle of **the conservation of energy** (when no work is done by nonconservative forces).

Example 7–1 A 0.25-kg box slides down a frictionless 2.1-m ramp that is inclined at 32° to the horizontal. Use energy considerations to determine its speed at the bottom if it starts from rest.

Setting it up:

The diagram shows the box sliding down the ramp. The information given in the problem is the following

Given: $m = 0.25$ kg, $L = 2.1$ m, $\theta = 32°$, $v_0 = 0$

Find: v

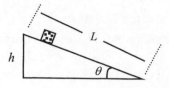

Strategy:

To solve this problem using the conservation of mechanical energy, we shall write expressions for both the initial and final energies and set them equal to each other. We set the reference for gravitational potential energy to be at the bottom of the ramp.

Working it out:

At the top, the total mechanical energy is

$$E_i = K_i + U_i = 0 + U_i = mgh = mgL\sin\theta.$$

At the bottom, the total mechanical energy is

$$E_f = K_f + U_f = \tfrac{1}{2}mv^2 + 0 = \tfrac{1}{2}mv^2.$$

We can now use the conservation of energy, $E_f = E_i$, to solve for v:

$$\tfrac{1}{2}mv^2 = mgL\sin\theta \quad \Rightarrow \quad v = \sqrt{2gL\sin\theta}.$$

Therefore,

$$v = \sqrt{2(9.8 \text{ m/s}^2)(2.1 \text{ m})\sin(32°)} = 4.7 \text{ m/s}.$$

What do you think? How would the working of the problem be different if the angle were at either of the extremes of $\theta = 0°$ or $90°$?

Example 7–2 At a party, a 0.63-kg ball is going to be shot vertically upward using a spring-loaded mechanism. The spring has a force constant of 188 N/m and is initially compressed, by both the ball and the person loading it, to 45 cm from equilibrium. **(a)** How high will the ball go above the compressed position and **(b)** what will its speed be at half of this height? (Assume the ball leaves contact with the spring when the spring reaches its usual equilibrium position.)

Setting it up:

The sketch on the left shows the ball loaded onto the spring, and the sketch on the right shows it after launch. The information given in the problem is the following

Given: $m = 0.63$ kg, $k = 188$ N/m, $y = 45$ cm, $v_0 = 0$

Find: **(a)** h, **(b)** v

Strategy:

To determine the height of the ball we can use energy conservation, since only conservative forces are doing work here. Set $U_{grav} = 0$ at the initial position of the ball.

Working it out:

(a) Initially, the ball is not moving ($K_0 = 0$) and we take $U_{grav} = 0$ there. So, the initial total energy is all potential energy due to the spring

$$E_0 = \tfrac{1}{2}ky^2 .$$

At the very top, the ball is again not moving. So, the total energy is all potential energy due to gravity

$$E_f = mgh .$$

We can set these equal and solve for the height,

$$mgh = \frac{1}{2}ky^2 \quad \Rightarrow \quad h = \frac{ky^2}{2mg} .$$

This then gives

$$h = \frac{(188 \text{ N/m})(0.45 \text{ m})^2}{2(0.63 \text{ kg})(9.8 \text{ m/s}^2)} = 3.1 \text{ m} .$$

(b) Write the total energy for the system when the ball is at half the calculated height in part (a)

$$E = \tfrac{1}{2}mv^2 + \tfrac{1}{2}mgh .$$

Setting this equal to the total energy, mgh, allows us to easily solve for the speed,

$$\tfrac{1}{2}mv^2 + \tfrac{1}{2}mgh = mgh \quad \Rightarrow \quad v = \sqrt{gh} .$$

Thus,

$$v = \sqrt{(9.8 \text{m/s}^2)(3.083 \text{m})} = 5.5 \text{m/s} .$$

What do you think? Would the answer to part (a) be greater or lesser if the spring were stiffer?

Practice Quiz

1. When an object is thrown up into the air, the gravitational potential energy
    (a) increases on the way up and decreases on the way down
    (b) decreases on the way up and increases on the way down
    (c) changes the same way going up as it does going down
    (d) remains constant throughout its motion
    (e) none of the above

2. The potential energy of a spring that is *stretched* an amount x from equilibrium versus one that is *compressed* an amount x from equilibrium
   (a) increases as it is stretched and decreases as it is compressed
   (b) decreases as it is stretched and increases as it is compressed
   (c) changes the same way whether the spring is stretched or compressed
   (d) remains constant throughout its motion
   (e) none of the above

3. A 2.0-kg object is lifted 2.0 m off the floor, carried horizontally across the room, and finally placed down on a table that is 1.5 m high. What is the overall change in gravitational potential energy for the Earth-object system?
   (a) 6.0 J   (b) 20 J   (c) 29 J   (d) 9.8 J   (e) 59 J

4. If a 0.75-kg block is rested on top of a vertical spring of force constant 45 N/m, how much potential energy is stored in the spring relative to its equilibrium configuration?
   (a) 0 J   (b) 14 J   (c) 60 J   (d) 34 J   (e) 0.60 J

5. If mechanical energy is conserved within a system,
   (a) all objects in the system move at constant speed
   (b) there is no change in the potential energy of the system
   (c) only gravitational forces act within the system
   (d) all the energy is in the form of potential energy
   (e) none of the above

6. A horizontal spring of force constant $k = 100$ N/m is compressed by 5.6 cm, placed against a 0.25-kg ball (that rests on a frictionless, horizontal surface), and then released. How fast will the ball be moving if it loses contact with the spring when the spring passes its equilibrium point?
   (a) 1.3 m/s   (b) 1.1 m/s   (c) 1.4 m/s   (d) 140 m/s   (e) 2.5 m/s

7–2 — 7–3 Energy Conservation and Allowed Motion and Motion in Two and Three Dimensions

A plot showing how the energy of a particle changes with position is called an *energy diagram*. Such a diagram can provide useful information about the allowed motions of the system. Consider the following diagram for a case in which the mechanical energy is conserved.

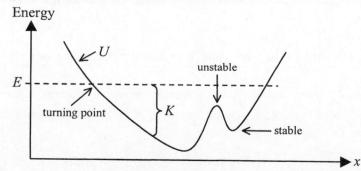

The dashed horizontal line represents the value of the total mechanical energy of the system. The difference between the potential energy and the mechanical energy must equal the kinetic energy, K. Where the horizontal line intersects the curve is where the energy of the system is completely in the form of potential energy, so that $K = 0$; the motion of the system must stop there. These points are called **turning points** because objects within the system stop moving in one direction and begin moving in the opposite direction.

Within the *potential well* between the two turning points, the motion will be periodic, going back and forth between the two turning points. Regions of the curve that are higher than E would correspond to negative kinetic energy, which is not possible. These are the *forbidden regions* of the motion. Because the force on a particle in the system equals the negative of the slope of $U(x)$, points at which the slope is zero are points at which the system is in equilibrium. Local minima are points of **stable equilibrium** because the force tries to push the particle back toward the equilibrium point. Local maxima are points of **unstable equilibrium** because the force will push the particle further away from the equilibrium point.

Example 7–3 A ball, starting at point a, moves under the influence of a force that corresponds to the potential energy curve shown. Describe the final motion of the ball.

Setting it up:
The diagram shows the potential energy curve labeled as U, the mechanical energy E and five specific points labeled $a \rightarrow e$.

Strategy/Working it out:
At point a, $U = E$, so the ball is at rest. It will gain kinetic energy as it moves toward b so its speed increases from a to b. As the ball moves from b to c, it loses kinetic energy as it gains potential energy, so it slows down; however, it does not lose all its kinetic energy so it does not stop at c. In moving from c to d it speeds up again, moving faster at d than it did at b. On its way to e, the ball gets slower and slower eventually coming to rest, momentarily at e.

After coming to rest at e, the ball's motion reverses, and it has the same speeds at the same points along its path. The ball eventually comes back to rest at point a. Since there is no work done to change the mechanical energy, this cycle continues with the ball going back and forth between the turning points a and e indefinitely.

What do you think? How might the motion be described if ball experiences a small frictional force?

The energy diagram shown above is of a one-dimensional potential energy function. In many situations, U is a function of three-dimensional space (x, y, z). In these cases it can be useful to plot just the lines of constant potential energy — the *equipotentials*. This type of analysis is more economical than using three components of force and is used widely in chemistry.

An important example of a force that depends on (x, y, z) is a **central force**. The functional dependence of a central force is such that its magnitude only depends on the combination $r = \sqrt{x^2 + y^2 + z^2}$, which is the radial distance from a fixed point. Furthermore, the direction of a central force is along this radial line,

$$\vec{F}(\vec{r}) = F(r)\hat{r} ,$$

where $\hat{r}$ is a unit vector pointing away from the fixed point. The two most important examples of central forces are gravity and the electrostatic force. All central forces are conservative. For this special case, the relationship between the force and the potential energy can be written in a way very much as in the one-dimensional case

$$\vec{F} = -\frac{dU(r)}{dr}\hat{r} .$$

This situation will be discussed further in the discussions on gravity and electrostatics.

Practice Quiz

7. When an object reaches a turning point in its motion
    (a) it must stop and remain at rest
    (b) it must continue moving past this point
    (c) all of its mechanical energy is in the form of kinetic energy
    (d) none of its mechanical energy is in the form of kinetic energy
    (e) none of the above

7–4 Is Energy Conservation a General Principle?

In section 7-1, we established the conservation of total mechanical energy only for the case in which no work is done by nonconservative forces. So, is energy conserved when work is done by nonconservative forces such as friction? Strictly speaking, yes, energy is always conserved even in the presence of nonconservative forces. However, in the earlier discussion we only considered the total *mechanical* energy (sum of kinetic and potential energies). When friction is involved, we also must account for energy converted into other forms, particularly thermal energy. This energy is often difficult to keep track of accurately. So, even though $E (= K+U)$ is not conserved, when all forms of energy are considered, energy is conserved.

To handle situations involving nonconservative forces, we can divide the work done by the net force into work done separately by conservative and nonconservative forces

$$W_{net} = W_{nc} + W_{conserv} = \Delta K .$$

The work done by conservative force is written as a change in potential energy

$$W_{nc} = \Delta K + \Delta U = \Delta E .$$

So, the work done by nonconservative forces produces a change in the total mechanical energy of the system.

Example 7–4 In the 1998 Winter Olympic games in Nagano, Japan, Georg Hackl of Germany won the gold medal in the luge. The total course length was 1326 m with a vertical drop of 114 m. If Georg made a starting leap that gave him an initial speed of 1.00 m/s, and the coefficient of kinetic friction was 0.0222, what was his speed at the finish line?

Setting it up:

The luge track is a winding and uneven surface, so a straight inclined plane would not be an accurate picture. However, on average, the results work out to be the same as you would get on a straight inclined plane. Thus, we work the problem as if it were a straight inclined path with the understanding that intermediate results can be considered to be only average values. We can reduce the problem to the following information:

Given: $L = 1326$ m, $\Delta y = -114$ m, $v_i = 1.00$ m/s, $\mu_k = 0.0222$;   **Find**: v_f

Strategy:

In this problem, we cannot apply the conservation of mechanical energy because work is being done by kinetic friction (a nonconservative force), and the amount of work done by friction causes a change in the total mechanical energy according to the equation $W_f = \Delta E$.

Working it out:

The work done by friction is

$$W_f = -f_k L = -\mu_k N L = -\mu_k mg \cos(\alpha)L ,$$

where the angle is $\alpha = \sin^{-1}(|\Delta y|/L) = 4.932°$ and the minus sign is from $\cos(180°)$ because the force of friction is in the opposite direction of the displacement. If we take the potential energy to be zero at the bottom of the track, the change in mechanical energy is

$$\Delta E = \Delta K + \Delta U = \tfrac{1}{2}mv_f^2 - \tfrac{1}{2}mv_i^2 + mg\Delta y$$

Putting it all together, we have

$$-\mu_k mgL \cos\alpha = \tfrac{1}{2}mv_f^2 - \tfrac{1}{2}mv_i^2 + mg\Delta y$$

Solving for v_f gives

$$v_f = \left[v_i^2 - 2g(\Delta y) - 2\mu_k gL \cos\alpha \right]^{1/2}$$

$$= \left[(1.00\,\text{m/s})^2 - 2(9.80\ \text{m/s}^2)(-114\,\text{m}) - 2(0.0222)(9.80\ \text{m/s}^2)(1326\,\text{m})\cos(4.932°) \right]^{1/2} = 40.8\ \text{m/s}$$

In this problem, every term requires knowledge of Georg's mass, which was not given. Precisely because every term required it, we were able to cancel it in the equation. Don't necessarily think you're stuck if you don't know something that appears to be required; it may cancel out in the long run.

What do you think? Would the amount of time to lower the weight back to the original location at constant speed be any different?

Practice Quiz

8. Which of the following statements is most accurate?

   (a) Mechanical energy is always conserved.

   (b) Mechanical energy is conserved only when potential energy is transformed into kinetic energy.

   (c) Mechanical energy is conserved when no nonconservative forces are applied.

   (d) Mechanical energy is conserved when no work is done by nonconservative forces.

   (e) Mechanical energy is never conserved.

9. Work done by a nonconservative force

   (a) can change only the kinetic energy

   (b) can change only the potential energy

   (c) can change either the kinetic or potential energy, or both

   (d) Work cannot be done by a nonconservative force.

   (e) none of the above

10. In which of the following cases does work done by a nonconservative force cause a change in potential energy?

    (a) A spring launches a ball into the air.

    (b) A person throws a ball into the air.

    (c) A ball falls vertically downward with negligible air resistance.

    (d) A sliding block on a horizontal surface comes to rest.

    (e) none of the above

11. A block given an initial speed of 5.7 m/s comes to rest due to friction after sliding 2.4 m upward along a surface that is inclined at 30 degrees. What is the coefficient of kinetic friction?

    (a) 0.22   **(b)** 0.42   **(c)** 0.58   **(d)** 0.50   **(e)** 0.87

Reference Tools and Resources

I. Key Terms and Phrases

potential energy an energy associated with conservative forces related to the configuration of a system

total mechanical energy the sum of the kinetic and potential energies in a system

the conservation of energy the principle that the total energy of system remains constant

turning point a point in an object's motion at which it will stop and reverse direction

stable equilibrium a point of equilibrium about which the applied force pushes the system back toward equilibrium

unstable equilibrium a point of equilibrium about which the applied force pushes the system further away from equilibrium

central force A force for which the magnitude depends only on the radial distance from a fixed point and the direction is along the radial line from this point

II. Important Equations

Name/Topic	Equation	Explanation
Potential energy	$\Delta U = -W_{AB}$	Potential energy is the negative of the work done by a conservative force.
Potential energy for gravity	$U(y) = mgy + U_0$	The potential energy for gravity near Earth's surface.
Potential energy for a spring	$U(x) = \frac{1}{2}kx^2$	The potential energy for a spring assuming $U = 0$ for the undeformed configuration.
Total mechanical mnergy	$E = K + U$	The total mechanical energy is the sum of the kinetic and potential energies.
The conservation of energy	$E_0 = E_f$	The total mechanical energy is conserved when only conservative forces do work.
Energy with nonconservative forces	$W_{nc} = \Delta K + \Delta U = \Delta E$	In the presence of nonconservative forces, the total mechanical energy can change.

III. Know Your Units

Quantity	Dimension	SI Unit
Potential energy (U)	$\left[ML^2T^{-2} \right]$	J
Mechanical energy (E)	$\left[ML^2T^{-2} \right]$	J

Practice Problems

1. The block is falling at a speed of 19 m/s and is 10 meters above the spring. The spring constant is 4000 N/m. To the nearest tenth of a cm, how far will the spring be compressed?

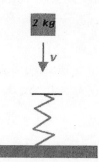

2. In the previous problem, to what height will the block rise after it hits and leaves the spring, to the nearest hundredth of a meter?

3. In the figure h_1 = 2.9 m and h_2 = 2.6 m. Use the floor as the zero of potential energy. To the nearest tenth of a joule, what is the potential energy of a 1-kg test mass placed at point A?

4. In the previous problem, taking the zero of potential energy at the ceiling, what is the potential energy of the test mass at point B?

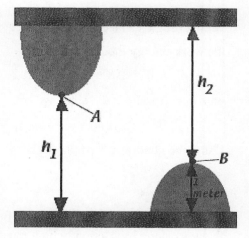

5. In Problem 3, use the floor as the zero of potential energy. To the nearest tenth of a joule, what is the difference in potential energy between points A and B?

6. In Problem 3, use the ceiling as the zero of potential energy. To the nearest tenth of a joule, what is the difference in potential energy between points A and B?

7. A 1-kg ball starting at h = 6 meters slides down a smooth surface where it encounters a rough surface and is brought to rest at B, a distance 21.1 meters away. To the nearest joule, what is the work done by friction?

8. In the previous problem, what is the coefficient of friction, to two decimal places?

9. The potential energy of a particle traveling along the x axis is given by $U = x^2 + 3.6x$. The total energy is 35 joules. To the nearest hundredth of a meter, where is the positive turning point?

10. The following potential energy has one stable point: $U = x^3 + 6x^2 + 12x + 36$. To two decimal places, where is it?

Selected Solutions to End of Chapter Problems

11. Because the energy is conserved, we have

$$E = K_1 + U_1 = K_2 + U_2 \quad \Rightarrow \quad \tfrac{1}{2}mv_1{}^2 + \tfrac{1}{2}kx_1{}^2 = \tfrac{1}{2}mv_2{}^2 + \tfrac{1}{2}kx_2{}^2$$

Before the block hits the spring its potential energy is zero, $U_1 = 0$. Once the block stops, it kinetic energy is zero, $K_2 = 0$. So, the above equation becomes

$$\tfrac{1}{2}mv_1^2 = \tfrac{1}{2}kx_2^2 \quad \Rightarrow \quad x_2 = \left(mv_1^2/k\right)^{1/2}$$

$$x_2 = \left(\frac{0.528\ \text{kg}\,(3.85\ \text{m/s})^2}{26.7\ \text{N/m}}\right)^{1/2} = \boxed{0.541\ \text{m}}$$

With a rough surface, there will be work done by the friction force, $f = -\mu_k N$, with $N = mg$. Thus, by the work-energy theorem

$$W_f = \Delta K + \Delta U \quad \Rightarrow \quad -\mu_k mg\,x = (\tfrac{1}{2}mv_2{}^2 - \tfrac{1}{2}mv_1{}^2) + (\tfrac{1}{2}kx_2{}^2 - \tfrac{1}{2}kx_1{}^2)$$

with $v_2 = 0$ and $x_1 = 0$. Thus, we get

$$-\mu_k mgx = -\tfrac{1}{2}mv_1^2 + \tfrac{1}{2}kx_2^2 \quad \Rightarrow \quad (k/2)x^2 + \left(\mu_k mg\right)x - \tfrac{1}{2}mv_1^2 = 0$$

Using the quadratic formula gives

$$x = \frac{-\mu_k mg \pm \sqrt{\left(\mu_k mg\right)^2 + mkv_1^2}}{k}$$

The solutions of the quadratic equation are $x = 0.468$ m, and $x = -0.627$ m. From our expression for the work done by friction, which must be negative, we select the positive result: $x = \boxed{0.468\ \text{m}}$.

13. We choose $y = 0$ at the relaxed position of the spring and denote the height of release by h and the magnitude of the compression of the spring by Δy. Because the energy is conserved, we have

$$E = K + U_g + U_{\text{spring}} = \text{constant}$$

Initially and finally, the ball is at rest, so $K = 0$ both at the beginning and the end.

$$0 + mgh + 0 = 0 + mg(-\Delta y) + \tfrac{1}{2}k(-\Delta y)^2$$

This gives the following quadratic equation for Δy:

$$\frac{k}{2}(\Delta y)^2 - mg\Delta y - mgh = 0$$

The two solutions of this quadratic equation are $\Delta y = 0.43$ m, and $\Delta y = -0.41$ m. From our choice of Δy as a magnitude it must be positive, we therefore select the positive value: $\Delta y = \boxed{0.43\ \text{m}}$.

For the man, the same quadratic equation applies; the only difference is the value of the mass. The resulting solution yields $\Delta y = \boxed{0.48\ \text{m}}$. The greater decrease in gravitational potential energy requires a greater spring potential energy.

27. The potential energy is $U(x) = \alpha x^4 + \beta x^2$. A sketch of this function is shown. The equilibrium positions have $dU/dx = 0$:

$dU/dx = 4\alpha x^3 + 2\beta x = 0$

from which we get $x = 0$ m, and

$x = \pm(-\frac{1}{2}\beta/\alpha)^{1/2} = \pm[-\frac{1}{2}(-3.0 \text{ J/m}^2)/(26 \text{ J/m}^4)]^{1/2} = \pm 0.24$ m.

From the sketch, we see that $\boxed{x = 0 \text{ is unstable and } x = \pm 0.24 \text{ m is stable}}$.

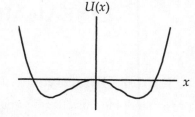

37. With resistive forces present we use the work-energy theorem, giving

$$W_f = \Delta K + \Delta U = (\tfrac{1}{2}mv_C^2 - \tfrac{1}{2}mv_A^2) + mg(h_C - h_A)$$
$$= [\tfrac{1}{2}(75 \text{ kg})(23 \text{ m/s})^2 - 0] + (75 \text{ kg})(9.8 \text{ m/s}^2)(0 - 95 \text{ m}) = \boxed{-5.0 \times 10^4 \text{ J}}.$$

47. The work done by the nonconservative drag forces changes the energy of the system:

$$W_{nc} = \Delta(K + U) = (K_f + U_f) - (K_i + U_i) = (\tfrac{1}{2}mv_f^2 + 0) - (0 + mgh_i)$$
$$= \tfrac{1}{2}(75 \text{ kg})(5.0 \text{ m/s})^2 - (75 \text{ kg})(9.8 \text{ m/s}^2)(85 \text{ m}) = \boxed{-6.2 \times 10^4 \text{ J}}.$$

49. We assume one 75-W bulb in each room that is kept lit for 2 hours in the morning and 5 hours in the evening each day:

$$\text{Cost} = 3(75 \text{ W})(2 \text{ h} + 5 \text{ h})(1 \text{ kW}/1000 \text{ W})(12 \text{ cents/kWh}) = \boxed{20 \text{ cents}}.$$

Answers to Practice Quiz
1. (a) 2. (c) 3. (c) 4. (e) 5. (e) 6. (b) 7. (d) 8. (d) 9. (c) 10. (b) 11. (a)

Answers to Practice Problems

1.	52.8 cm	2.	28.40 m
3.	28.4 J	4.	−25.5 J
5.	18.6 J	6.	18.6 J
7.	−59 J	8.	0.28
9.	4.38 m	10.	−2.00

CHAPTER 8

LINEAR MOMENTUM, COLLISIONS, AND THE CENTER OF MASS

Chapter Objectives

After studying this chapter, you should

1. be able to use Newton's second law in terms of linear momentum.
2. be able to apply the relationship between impulse and linear momentum.
3. be able to use the conservation of momentum to analyze elastic and inelastic collisions.
4. be able to determine the center of mass of systems of discrete particles and certain continuous distributions of matter.
5. be able to apply Newton's laws to the motion of the center of mass of a system.
6. be able to determine the speed of a rocket after a certain amount of fuel has been expelled.

Chapter Review

In Chapters 6 and 7, we saw that defining concepts beyond just force, velocity, acceleration, and displacement made analyzing some situations much easier, especially by using the conservation of energy. In this chapter, we do more of the same by introducing the concept of linear momentum. In a way similar to work and energy, the idea of linear momentum often provides a more direct path to understanding physical interactions than working with forces only. Furthermore, as with energy, there is a conservation principle for linear momentum that is of tremendous value in physics.

8–1 Momentum and Its Conservation

The form of Newton's second law that we have been using until now, $\vec{F}_{net} = m\vec{a}$, applies only to circumstances in which the mass remains constant; however, in the most general cases, as with rockets, the mass may change during the motion. The most general form of Newton's second law is

$$\vec{F}_{net} = \frac{d(m\vec{v})}{dt},$$

where the quantity being differentiated is called the linear momentum (often the word "linear" is not stated)

$$\vec{p} = m\vec{v}.$$

Linear momentum is a vector quantity whose direction is that of the velocity $\vec{v}$ and whose units are just the mass unit times velocity units, kg·m/s. Therefore, the best overall form of Newton's second law is

$$\vec{F}_{net} = \frac{d\vec{p}}{dt}.$$

In addition to the net force on an object, the kinetic energy of an object is also conveniently expressed in terms of its linear momentum as

$$K = \frac{p^2}{2m}.$$

For a system of particles, we often speak of the *total linear momentum* of the system. This total linear momentum is the vector sum of the momentum of each particle in the system:

$$\vec{P}_{total} = \vec{p}_1 + \vec{p}_2 + \vec{p}_3 + \cdots$$

If we consider only interactions between the particles in the system, that is, for an isolated system, Newton's second and third laws of motion tell us that the total linear momentum of such a system is constant. This fact is called the principle of the **conservation of momentum**. This principle is one of the most important principles in physics.

Example 8–1 A 0.25-kg object moves due east at 2.1 m/s. What average force is needed to cause it to move due north at 3.6 m/s in 1.52 s?

Setting it up: The information given in the problem is the following
Given: $m = 0.25$ kg, $\vec{v}_i = 2.1$ m/s $\hat{i}$, $\vec{v}_f = 3.6$ m/s $\hat{j}$, $t = 1.52$ s;   **Find**: $\vec{F}$

Strategy:
Even though this is a case of constant mass, let's work it in terms of momentum to illustrate that approach.

Working it out:
The initial and final momenta of the object are

$$\vec{p}_i = m\vec{v}_i = 0.25 \text{ kg}(2.1 \text{ m/s})\hat{i} = 0.525 \text{ kg}\cdot\text{m/s } \hat{i}$$
$$\vec{p}_f = m\vec{v}_f = 0.25 \text{ kg}(3.6 \text{ m/s})\hat{j} = 0.900 \text{ kg}\cdot\text{m/s }\hat{j}.$$

The change in momentum, therefore, is

$$\Delta\vec{p} = \vec{p}_f - \vec{p}_i = \left(-0.525\hat{i} + 0.900\,\hat{j}\right)\text{kg}\cdot\text{m/s}.$$

The force, then, is given by

$$\vec{F}_{av} = \frac{\Delta\vec{p}}{\Delta t} = \frac{\left(-0.525\,\hat{i} + 0.900\,\hat{j}\right)\text{kg}\cdot\text{m/s}}{1.52 \text{ s}} = \left(-0.35\,\hat{i} + 0.59\,\hat{j}\right)\text{N}$$

What do you think? Solve this problem using $\vec{F}_{av} = m\vec{a}_{av}$ to see if you get the same result.

Practice Quiz

1. Two objects have equal velocities but one has twice the mass of the other. The object with larger mass

   (a) has half the momentum of the other
   (b) has twice the momentum of the other

(c) has the same momentum as the other

(d) must move slower than the other

(e) none of the above

2. If two objects have the same mass and speed but move in opposite directions, then
    (a) they have momenta of equal magnitude and opposite directions
    (b) they have equal momenta
    (c) each has zero momentum
    (d) one must move faster than the other
    (e) none of the above

3. When does the equation $\vec{F} = m\vec{a}$ not accurately describe the dynamics of an object?
    (a) when gravity is not present
    (b) when more than one force acts on the object
    (c) when kinetic friction does work on the object
    (d) when the mass of the object is constant
    (e) none of the above

4. If the linear momentum of a 1.6-kg object is decreasing at a rate of 5.0 $\frac{\text{kg·m/s}}{\text{s}}$, what is the magnitude of the force on the object?
    (a) 8.0 N   (b) 3.4 N   (c) 5.0 N   (d) 6.6 N   (e) 3.1 N

5. Which of the following statements is most accurate?
    (a) The linear momentum of a system is always conserved.
    (b) The linear momentum of a system is conserved only if no external forces act on the system.
    (c) The linear momentum of a system is conserved if the net external force on the system is zero.
    (d) The linear momentum of a system is not conserved.
    (e) None of the above.

8–2 Collisions and Impulse

The concept of momentum is important when two objects interact. In this chapter, the interaction we focus on is called a **collision**. A collision occurs when the forces of interaction between two objects are large for a finite period of time; in a situation like this we say that the force is an **impulsive force**. From Newton's second law, $\vec{F} = d\vec{p}/dt$, we can see that the result of a force acting over a short period of time, $\vec{F}dt$, is a change in the linear momentum $d\vec{p}$. The net effect over the time period that the force acts is called the impulse of the force $\vec{J}$

$$\vec{J} = \int_{t_0}^{t_f} \vec{F} dt = \Delta \vec{p} .$$

It is also useful to note that the impulse can be determined from the average force that is applied over the time interval

$$\vec{J} = \vec{F}_{av}\,\Delta t\,.$$

The expression $\vec{J} = \Delta \vec{p}$ is referred to by many authors as the impulse-momentum theorem.

Example 8–2 The velocity of a rock of mass 0.24 kg moving with a speed of 3.33 m/s makes a 60° angle with the normal to a brick wall. The rock is in contact with the wall for only 0.032 s. If the velocity of the rock makes an angle of 40° with the normal to the wall after it strikes and has a magnitude of 2.68 m/s, **(a)** what impulse does the wall apply to the rock, and **(b)** what average force causes this impulse?

Setting it up:

A diagram is helpful here. The diagram shows the initial and final momenta of the rock as it bounces off the wall. The information given in the problem is the following:

Given: $m = 0.24$ kg, $v_i = 3.33$ m/s, $\theta_i = 60°$, $v_f = 2.68$ m/s,

$\qquad \theta_f = 40°$

Find: **(a)** $\vec{J}$, **(b)** $\vec{F}_{av}$

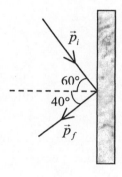

Strategy:

To solve for the impulse, we can find the change in momentum that results from the bounce. Once we know the impulse, we'll use it to get the average force.

Working it out:

(a) First, we can determine the components of the initial and final momenta of the rock

$$p_{i,x} = mv_i \cos(60°) = 0.24 \text{ kg}(3.33 \text{ m/s})\cos(60°) = 0.3996 \text{ kg} \cdot \text{m/s}$$

$$p_{i,y} = -mv_i \sin(60°) = -0.24 \text{ kg}(3.33 \text{ m/s})\sin(60°) = -0.6921 \text{ kg} \cdot \text{m/s}$$

$$p_{f,x} = -mv_f \cos(40°) = -0.24 \text{ kg}(2.68 \text{ m/s})\cos(40°) = -0.4927 \text{ kg} \cdot \text{m/s}$$

$$p_{f,y} = -mv_f \sin(40°) = -0.24 \text{ kg}(2.68 \text{ m/s})\sin(40°) = -0.4134 \text{ kg} \cdot \text{m/s}$$

Now, the impulse can be found as the change in momentum:

$$\vec{J} = \vec{p}_f - \vec{p}_i = (p_{f,x} - p_{i,x})\hat{i} + (p_{f,y} - p_{i,y})\hat{j} = [(-0.4927 - 0.3996)\hat{i} + (-0.4134 + 0.6921)\hat{j}]\,\text{kg} \cdot \text{m/s}$$

$$= \left(-0.892\,\hat{i} + 0.279\,\hat{j}\right)\text{N} \cdot \text{s}$$

(b) We can now use the relationship between force and impulse to obtain the average force

$$\vec{J} = \vec{F}_{av}\,\Delta t \quad \Rightarrow \quad \vec{F}_{av} = \vec{J}/\Delta t$$

Therefore, the numerical value of the force is

$$\vec{F}_{av} = \frac{\left(-0.892\,\hat{i} + 0.279\,\hat{j}\right)\text{N} \cdot \text{s}}{0.032 \text{ s}} = \left(-28\,\hat{i} + 8.7\,\hat{j}\right)\text{N}$$

What do you think? For what reason might the ball bounce off at a different angle than the incident?

The concept of impulse is important in collisions because when two objects collide they impart equal and opposite impulses on each other and therefore, equal and opposite changes in momentum. This fact is how Newton's laws of motion account for the conservation of linear momentum in collisions. In this chapter, we treat three classifications of collisions, **(1)** two objects collide and stick together; **(2)** the two colliding objects remain distinct and unchanged after the collision, and **(3)** some mass transfers from one object to the other. In each of these cases, we can apply both the conservation of linear momentum

$$\vec{P}_{\text{total, before}} = \vec{P}_{\text{total, after}},$$

and the conservation of mass

$$\sum_{\text{before}} m_i = \sum_{\text{after}} m_i.$$

In most collisions, some energy is dissipated to generate heat, sound, and material deformations. There is, however, a special category of collisions in which these energy losses do not occur. In such cases, the total kinetic energy of the system before and after the collision is the same; we call this type an **elastic collision**. All other cases are called **inelastic collisions**. The special case in which the colliding bodies stick together, case 1 above, corresponds to the maximum loss of kinetic energy and is called a **perfectly inelastic collision**.

Practice Quiz

6. During one trial, a force F is applied to a ball for an amount of time t. During a second trial a force of $3F$ is applied for a time of $t/2$. Which of the following is true concerning the magnitude of the change in momentum of the ball?

(a) There's a greater change for the second trial.

(b) There's a smaller change for the second trial.

(c) We get the same nonzero momentum change in each trial.

(d) The momentum doesn't change in either trial.

(e) None of the above.

7. An object of mass 2.8 kg moves at 1.1 m/s in the $+x$ direction. If an impulse of $(1.3 \, \text{N} \cdot \text{s}) \, \hat{i}$ is applied, what is its final momentum?

(a) 3.1 kg·m/s   (b) 4.0 kg·m/s   (c) 5.2 kg·m/s   (d) 4.4 kg·m/s   (e) 4.2 kg·m/s

8–3 Perfectly Inelastic Collisions; Explosions

In a perfectly inelastic collision between two bodies, the final state is that all of the momentum is contained in the single body produced when the objects stick together. The equations we use to analyze such a collision are

$$m_1\vec{v}_1 + m_2\vec{v}_2 = M\vec{v} \quad \text{and} \quad M = m_1 + m_2,$$

where the first equation represents the conservation of momentum and the second represents the conservation of mass. The amount to kinetic energy lost in perfectly inelastic collisions can be easily calculated. The result is

$$\Delta E = K_f - K_0 = \tfrac{1}{2}Mv^2 - \left(\tfrac{1}{2}m_1v_1^2 + \tfrac{1}{2}m_2v_2^2\right) = -\frac{m_1m_2\left(v_1 - v_2\right)^2}{2M}.$$

An explosion is the reverse of a completely inelastic collision. In an explosion, a single object becomes two or more. However, we still have the conservation of momentum and the conservation of mass (in macroscopic explosions) to aid the analysis. In an explosion, the kinetic energy of the final products comes from potential energy stored in the initial object.

Example 8–3 The driver of a 1485-kg car is driving along a street at 25 mi/h. Another driver, coming from directly behind (and not paying attention), is driving a 1520-kg car at 38 mi/h and hits the slower car. If both drivers slam the brakes at the moment of impact and their fenders catch on each other, with what combined speed do they begin to skid?

Setting it up:
The following information is given in the problem:
Given: $m_1 = 1485$ kg, $v_1 = 25$ mi/h, $m_2 = 1520$ kg, $v_2 = 38$ mi/h,   **Find:** v

Strategy:
This is a perfectly inelastic collision. We apply the conservation of momentum by setting the total momenta before and after the collision equal to each other. Take the direction of motion to be the x direction; we will, therefore, only work with the x components of the quantities.

Working it out:
The total momentum before the collision can be written as

$$p_{total} = m_1v_1 + m_2v_2.$$

The total momentum after the collision can be written as

$$p_{total} = (m_1 + m_2)v.$$

Because momentum is conserved, we set them equal to each other

$$m_1v_1 + m_2v_2 = (m_1 + m_2)v \quad \therefore \quad v = \frac{m_1v_1 + m_2v_2}{m_1 + m_2}.$$

The final result then is

$$v = \frac{1520\text{ kg}\left(38\tfrac{\text{mi}}{\text{h}}\right) + 1485\text{ kg}\left(25\tfrac{\text{mi}}{\text{h}}\right)}{1520\text{ kg} + 1485\text{ kg}} = 32 \text{ mi/h}.$$

What do you think? What would the result be if the cars were traveling in opposite directions?

Practice Quiz

8. A cart is rolling along a horizontal floor when a box is dropped vertically into it. After catching the box, the cart

    (a) will move more slowly

    (b) will move more quickly

    (c) will move at the same speed as before

    (d) will move in a different direction

    (e) will come to a stop as a result of catching the box.

9. Two objects of equal mass of 1.7 kg move directly toward each other with equal speed of 2.2 m/s. If they stick together after the collision their speed will be

    (a) 0.65 m/s   **(b)** 2.2 m/s   **(c)** 1.1 m/s   **(d)** 3.7 m/s   **(e)** 0 m/s

8–4—8–5 Elastic Collisions in One, Two, and Three Dimensions

An elastic collision will be one dimensional if it is a head-on collision. In this case, the velocities of the objects after the collision (v_1' and v_2') are along the same line of motion as their velocities before the collision (v_1 and v_2). The equations for the conservation of linear momentum,

$$m_1 v_1 + m_2 v_2 = m_1 v_1' + m_2 v_2',$$

and kinetic energy,

$$\tfrac{1}{2} m_1 v_1^2 + \tfrac{1}{2} m_2 v_2^2 = \tfrac{1}{2} m_1 v_1'^2 + \tfrac{1}{2} m_2 v_2'^2,$$

can be combined to give several useful results.

One of these useful results is that the relative velocity of the objects before the collision is equal and opposite to the relative velocity after the collision,

$$v_1 - v_2 = v_2' - v_1'.$$

We can also consider the momentum and kinetic energy equations as a system of two simultaneous equations with the final velocities as the two unknowns. Solving for these final velocities yields

$$v_1' = \frac{m_1 - m_2}{m_1 + m_2} v_1 + \frac{m_2 - m_1}{m_1 + m_2} v_2,$$

and

$$v_2' = \frac{2m_1}{m_1 + m_2} v_1 + \frac{m_2 - m_1}{m_1 + m_2} v_2.$$

Example 8–4 A 1.3-kg ball initially moving with a speed of 3.1 m/s strikes a stationary 2.2-kg ball head-on. What are the final velocities of the two balls if the collision is elastic?

Setting it up:

We can reduce the problem to the following information:

Given: $m_1 = 1.3$ kg, $v_1 = 3.1$ m/s, $m_2 = 2.2$ kg, $v_2 = 0$;   **Find:** v_1' and v_2'

Strategy:

The situation described in the problem is precisely the situation just discussed above. We can make direct use of the equations obtained.

Working it out:

The result for the final speed of mass 1 is

$$v_1' = \frac{m_1 - m_2}{m_1 + m_2}v_1 + 0 = \frac{1.3 \text{ kg} - 2.2 \text{ kg}}{1.3 \text{ kg} + 2.2 \text{ kg}}(3.1 \text{ m/s}) = -0.80 \text{ m/s},$$

The result for the final speed of mass 2 is

$$v_2' = \frac{2m_1}{m_1 + m_2}v_1 + 0 = \frac{2(1.3 \text{ kg})}{1.3 \text{ kg} + 2.2 \text{ kg}}(3.1 \text{ m/s}) = 2.3 \text{ m/s}$$

What do you think? Is the final result consistent with the above statements about the relative velocity before and after an elastic collision?

For cases in which the collision does not occur head-on, that is, for glancing collisions, the problem must be treated in at least two dimensions. For a case such as this, the glancing blow can occur many different ways from nearly head-on to a near miss. Therefore, there is no unique solution that applies to all glancing collisions. However, the conservation of linear momentum and kinetic energy still tell us a lot about what can and cannot happen in these collisions. For the special case in which the masses of the colliding objects are equal, and one of them is initially at rest, we find that the final velocities of the bodies are perpendicular to each other $\vec{v}_1' \cdot \vec{v}_2' = 0$. This fact also implies that

$$v_1^2 = v_1'^2 + v_2'^2.$$

Example 8–5 Two objects, one (m_2) of mass 1.50 kg moving due south at 15.3 m/s and the other (m_1) of mass 1.19 kg moving northeast at 11.7 m/s, collide elastically. After they collide, m_2 moves with a velocity of 12.3 m/s at 10.9° north of east. What is the final velocity of m_1?

Setting it up:

A diagram is very helpful here. The diagram shows the two masses moving toward each other before colliding as well as the speed and direction of m_2 afterward. We can reduce the problem to the following information:

Given: $m_1 = 1.19$ kg, $v_1 = 11.7$ m/s, $\theta_1 = 45°$,
$m_2 = 1.50$ kg, $v_2 = 15.3$ m/s, $\theta_2 = -90°$,
$v_2' = 12.3$ m/s, $\theta_2' = 10.9°$; **Find:** v_1', θ_1'

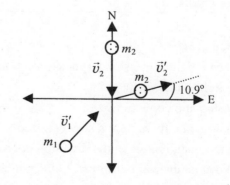

Strategy:

To perform the analysis we use the fact that kinetic energy is conserved together with the conservation of momentum in both the x and y directions.

Working it out:

The conservation of kinetic energy can be written as

$$K_{total}^{before} = K_{total}^{after} \quad \Rightarrow \quad \tfrac{1}{2}m_1v_1^2 + \tfrac{1}{2}m_2v_2^2 = \tfrac{1}{2}m_1v_1'^2 + \tfrac{1}{2}m_2v_2'^2,$$

$$\therefore \ m_1v_1^2 + m_2v_2^2 = m_1v_1'^2 + m_2v_2'^2$$

We can insert the numerical values of the quantities and solve this for v_1'

$$514.0 \text{ J} = (1.19 \text{ kg})v_1'^2 + 226.9 \text{ J} \quad \Rightarrow \quad v_1' = \sqrt{\frac{287.1 \text{ J}}{1.19 \text{ kg}}} = 15.5 \text{ m/s}.$$

Applying the conservation of momentum along the x direction gives

$$p_{total,x}^{before} = p_{total,x}^{after} \quad \Rightarrow \quad m_1v_1\cos(45°) = p_{1,x}' + m_2v_2'\cos(10.9°)$$

$$\therefore \ p_{1,x}' = m_1v_1\cos(45°) - m_2v_2'\cos(10.9°).$$

This gives

$$p_{1,x}' = 9.845 \text{ kg}\cdot\text{m/s} - 18.12 \text{ kg}\cdot\text{m/s} = -8.272 \text{ kg}\cdot\text{m/s}.$$

Applying the conservation of momentum along the y direction gives,

$$p_{total,y}^{before} = p_{total,y}^{after} \quad \Rightarrow \quad m_1v_1\sin(45°) - m_2v_2 = p_{1,y}' + m_2v_2'\sin(10.9°)$$

$$\therefore \ p_{1,y}' = (9.845 - 22.95 - 3.489) \text{ kg}\cdot\text{m/s} = -16.59 \text{ kg}\cdot\text{m/s}.$$

Now we can determine a direction for the final velocity of mass 1.

$$\theta_1' = \tan^{-1}\left(\frac{p_{1,y}'}{p_{1,x}'}\right) = \tan^{-1}\left(\frac{-16.59 \text{ kg}\cdot\frac{\text{m}}{\text{s}}}{-8.272 \text{ kg}\cdot\frac{\text{m}}{\text{s}}}\right) = 63.5°.$$

Recognizing that both components are negative, this angle is measured from the $-x$ axis. Therefore, we can say that v_1' is 15.5 m/s at 63.5° south of west.

What do you think?   Are the final velocities perpendicular to each other? If not, is this inconsistent with the prior discussion?

Practice Quiz

10. When a lighter mass undergoes a head-on, elastic collision with a heavier mass that is initially at rest, the lighter mass will

(a) stop and come to rest

(b) continue forward at a slower speed

(c) recoil and move in the direction opposite its original motion

(d) cannot answer definitively; it depends on the masses and the initial speed of the lighter mass

(e) none of the above

8–6 Center of Mass

At many times in our study we have treated large objects such as balls, cars, rockets, and so on, as if they were point particles even though they clearly are not. The reason we have been able to do that is because in many situations systems of particles behave (as a whole) just like point particles. The concept that helps us see this fact most clearly is the **center of mass**.

The center of mass of a system is the average location of mass in that system. This average is a weighted average in that each particle contributes more or less to the average according to its mass. The position of the center of mass, $\vec{R}$, can be determined from the positions of the individual particles $\vec{r}$ by the following expression:

$$\vec{R} = \frac{m_1\vec{r}_1 + m_2\vec{r}_2 + \cdots + m_N\vec{r}_N}{M},$$

where M is the total mass of the system, $M = m_1 + m_2 + \cdots + m_N$. The Cartesian components of this position vector are given by

$$X = \frac{m_1 x_1 + m_2 x_2 + \cdots + m_N x_N}{m_1 + m_2 + \cdots} = \frac{\sum m_i x_i}{M}$$

$$Y = \frac{m_1 y_1 + m_2 y_2 + \cdots + m_N y_N}{m_1 + m_2 + \cdots} = \frac{\sum m_i y_i}{M}$$

$$Z = \frac{m_1 z_1 + m_2 z_2 + \cdots + m_N z_N}{m_1 + m_2 + \cdots} = \frac{\sum m_i z_i}{M}.$$

Be aware that the center of mass is just an average location; there does not need to be any mass at that location. For a uniform circular ring, the center of mass is right at the center of the ring, where no mass is located. In fact, we can take it as given that for uniform, symmetric, and continuous distributions of matter, the center of mass is located at the geometric center of the system.

Example 8–6 Three objects are located at the following positions in a two-dimensional (x, y) coordinate system: m_1 at (1.3 m, 5.4 m), m_2 at (–2.2 m, 9.4 m), and m_3 at (4.1 m, –0.77 m). If the masses are 10 kg, 15 kg, and 20 kg respectively, what is the position of the center of mass of this system?

Setting it up:

This is a two-dimensional problem. We can reduce the problem to the following information:

Given: $m_1 = 10$ kg, $m_2, = 15$ kg, $m_3 = 20$ kg, $x_1 = 1.3$ m, $y_1 = 5.4$ m, $x_2 = -2.2$ m, $y_2 = 9.4$ m, $x_3 = 4.1$ m, $y_3 = -0.77$ m;

Find: X, Y

Strategy:

We can make direct use of the expression for calculating the center of mass.

Working it out:

$$X = \frac{\sum m_i x_i}{M} = \frac{m_1 x_1 + m_2 x_2 + m_3 x_3}{m_1 + m_2 + m_3} = \frac{10 \text{ kg}(1.3 \text{ m}) + 15 \text{ kg}(-2.2 \text{ m}) + 20 \text{ kg}(4.1 \text{ m})}{45 \text{ kg}} = 1.4 \text{ m}$$

$$Y = \frac{\sum m_i y_i}{M} = \frac{m_1 y_1 + m_2 y_2 + m_3 y_3}{m_1 + m_2 + m_3} = \frac{10 \text{ kg}(5.4 \text{ m}) + 15 \text{ kg}(9.4 \text{ m}) + 20 \text{ kg}(-0.77 \text{ m})}{45 \text{ kg}} = 4.0 \text{ m}$$

What do you think?   To which of the masses is the center of mass closest? Does the answer make sense to you intuitively?

It is often very useful to track the center of mass of a system of particles, because we can often describe the important aspects of the behavior of the system by treating the system as if it were a single particle with all of its mass located at the center of mass – this is especially true for rigid bodies. The velocity of the center of mass of a system is

$$\vec{V} = \frac{d\vec{R}}{dt} = \vec{P}/M \,,$$

where $\vec{P}$ is the total linear momentum $\vec{P} = \Sigma m_i \vec{v}_i$. Knowing that, in the absence of external force, the total linear momentum of a system is zero, we can see that this implies that the center of mass of such a system will move with constant velocity regardless of how complicated the motions of the individual parts of the system may be. This behavior is exactly what a single particle does when no net force is applied to it.

If there are external forces acting on the system, then these forces will affect each particle in the system according to Newton's second law, $\vec{F}_{i,\text{ext}} = d\vec{p}_i/dt$. The total external force on the system is the sum of all the external forces on all the particles, $\vec{F}_{\text{tot,ext}} = \Sigma \vec{F}_{i,\text{ext}} = \Sigma (d\vec{p}_i/dt) = d/dt\left(\Sigma \vec{p}_i\right)$. Therefore,

$$\vec{F}_{\text{tot,ext}} = \frac{d\vec{P}}{dt} \,.$$

So, just as for a single particle, the net external force on a system of particles equals the rate of change in its momentum. Furthermore, since $\vec{P} = M\vec{V}$, we see that $d\vec{P}/dt = M(d\vec{V}/dt)$. Therefore,

$$\vec{F}_{\text{tot,ext}} = M\vec{A} \,,$$

where $\vec{A}$ is the acceleration of the center of mass of the system. Just as in the absence of external force, the system can be treated like a single particle with all of its mass located at the center of mass.

Calculating the center of mass of most common objects using the summation above would be impossible because most common objects are made of billions of particles. For continuous distributions of matter the summation needs to be replaced by an integration. For objects in which the mass is distributed along a thin line, the location of the center of mass is determined from either

$$X = \frac{1}{M}\int x\,dm, \quad Y = \frac{1}{M}\int y\,dm, \text{ or } \quad Z = \frac{1}{M}\int z\,dm,$$

where dm is written in terms of the linear density function

$$\lambda(x) = \frac{dm}{dx} \quad \Rightarrow \quad dm = \lambda(x)dx.$$

In two dimensions, we have

$$dm = \sigma(x, y)dx\,dy ,$$

where σ is the surface mass density. Finally, in three dimensions

$$dm = \rho(x, y, z)dx\,dy\,dz ,$$

where ρ is the volume density (or just the density) of the object.

In certain situations it is easier to find the center mass than using the above calculations. For objects that are symmetric about a plane, line, or point, the center of mass will lie in that plane, on that line, or at that point. For example, the center of mass of a uniform ball is located at the geometric center of the ball. To that end, some objects that are not obviously symmetric can be viewed as a combination of smaller objects that are individually symmetric. By locating the centers of mass of these parts, we can then calculate the location of the center of mass using the summation.

Example 8–7 Two uniform, thin rods are attached so as to form an "L" shape. The vertical rod has a length of 0.50 m and a mass of 0.45 kg, while the horizontal rod has a length of 0.30 m and a mass of 0.25 kg. locate the center of mass of this system.

Setting it up:
A diagram is helpful here. The diagram shows the two rods forming the "L" shape. We can reduce the problem to the following information:
Given: $l_v = 0.50$ m, $m_v = 0.45$ kg, $l_h = 0.30$ m, $m_h = 0.25$ kg
Find: X, Y

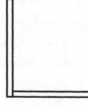

Strategy:
Because each rod is uniform, we can use symmetry to find their centers of mass. Once the coordinates of the centers of mass for each rod are specified, we can treat the system as two point particles.

Working it out:
The symmetry of the rods tells us that the center of mass of each rod is located at its geometric center. Taking the origin of our coordinate system at the bottom of the vertical rod with the positive y direction as up and the positive x direction to the right, we can identify the coordinates of the individual centers of mass as:

$$\text{vertical rod:} \quad X_v = 0,\ Y_v = 0.25 \text{ m}$$
$$\text{horizontal rod:} \quad X_h = 0.15 \text{ m},\ Y_h = 0 \text{ m.}$$

Now that the individual centers of mass have been located, we treat the system of rods as a system of particles. Thus,

$$X = \frac{\sum m_i X_i}{M} = \frac{m_v X_v + m_h X_h}{m_v + m_h} = \frac{0.45 \text{ kg}(0 \text{ m}) + 0.25 \text{ kg}(0.15 \text{ m})}{0.70 \text{ kg}} = 0.054 \text{ m}$$

$$Y = \frac{\sum m_i Y_i}{M} = \frac{m_v Y_v + m_h Y_h}{m_v + m_h} = \frac{0.45 \text{ kg}(0.25 \text{ m}) + 0.25 \text{ kg}(0 \text{ m})}{0.70 \text{ kg}} = 0.16 \text{ m} .$$

What do you think? Describe how you would perform this calculation with the integral approach alone. That is, without using symmetry and not treating it as a system of point particle.

Practice Quiz

11. For two objects of slightly different mass, the center of mass will be

   (a) just slightly away from the heavier mass, on the opposite side from the lighter one

   (b) between the two, just slightly closer to the lighter mass

   (c) very close to the lighter mass, on the opposite side from the heavier one

   (d) between the two, just slightly farther away from the lighter object

   (e) between the two, very close to the heavier mass

*8–7 Rocket Motion

As mentioned previously, Newton's second law in terms of momentum is a more general form for this law because it is not limited to systems with constant mass. A rocket is a perfect example of a system with changing mass because, at a typical launch, most of the rocket's mass is fuel, and the amount of this fuel decreases rapidly as it is exhausted. For a rocket whose exhaust is expelled backward at a speed u_{ex}, relative to the rocket, the forward force exerted on the rocket, called the **thrust**, is

$$F_{thrust} = -u_{ex} \frac{dm}{dt} ,$$

where dm/dt is the rate at which the mass of the rocket is changing and is negative.

Using this expression for the thrust, we can determine the speed of a rocket in terms of the amount of mass remaining as

$$v = u_{ex} \ln\left(\frac{m_0}{m}\right),$$

valid in "deep space" where there is no gravity. In the presence of gravity, the corresponding result is

$$v = u_{ex} \ln\left(\frac{m_0}{m}\right) - gt .$$

Example 8–8 At liftoff an advanced rocket has a total mass (payload + fuel) of 1.8×10^5 kg. If the rocket burns fuel at a constant rate of 3000 kg/s with an exhaust velocity of 4500 m/s, what is its acceleration 12.0 seconds after liftoff?

Setting it up:

We can reduce the problem to the following information:

Given: $m = 1.8 \times 10^5$ kg, $\Delta m / \Delta t = -3000$ kg/s, $u_{ex} = 4500$ m/s, $t = 12.0$ s;   **Find**: a

Strategy:

We need to determine the thrust that propels the rocket upward. The net force on the rocket from the thrust and gravity will then allow us to determine the acceleration at that instant.

Working it out:

The magnitude of the average thrust of the rocket is

$$F_{thrust} = \left| \frac{\Delta m}{\Delta t} \right| u_{ex} = (3000\,\text{kg/s})(4500\,\text{m/s}) = 1.35 \times 10^7\,\text{N}.$$

The mass of the rocket after 12.0 seconds is

$$m = m_0 - \left| \frac{\Delta m}{\Delta t} \right| t = 1.8 \times 10^5\,\text{kg} - (3000\,\text{kg/s})(12.0\,\text{s}) = 1.44 \times 10^5\,\text{kg}.$$

The net force on the rocket after 12.0 seconds is

$$F_{net} = F_{thrust} - mg = 1.35 \times 10^7\,\text{N} - (1.44 \times 10^5\,\text{kg})(9.80\,\text{m/s}^2) = 1.21 \times 10^7\,\text{N}.$$

Finally, the acceleration is

$$a = \frac{F_{net}}{m} = \frac{1.21 \times 10^7\,\text{N}}{1.44 \times 10^5\,\text{kg}} = 84\,\text{m/s}^2.$$

What do you think?   What is the estimated speed of this rocket 12.0 seconds after liftoff?

Practice Quiz

12. The force that thrusts a rocket forward comes from

   (a) pushing against the air

   (b) the pull of gravity

   (c) the reaction force from the exhaust

   (d) pushing against the ground at liftoff

   (e) none of the above

*8–8 Momentum Transfer at High Energies

At high energies, because the particles move very rapidly, interactions between these particles can take place at small distances. The ability to test the laws of physics at these small length scales has revealed, and is expected to continue to reveal, new and surprising aspects of nature.

Reference Tools and Resources

I. Key Terms and Phrases

conservation of momentum the principle that the total linear momentum of a system remains constant unless a nonzero external net force is applied

collision an interaction that involves forces that act appreciably for a finite period of time

impulsive force a force that is large for a short period of time

elastic collision a collision in which kinetic energy is conserved

inelastic collision a collision in which kinetic energy is not conserved

perfectly inelastic collision an inelastic collision in which the colliding bodies stick together

center of mass the weighted average location of the mass in a system

thrust the forward force on a rocket due to its exhaust

II. Important Equations

Name/Topic	Equation	Explanation
Linear momentum	$\vec{p} = m\vec{v}$	The definition of linear momentum.
Newton's second law	$\vec{F}_{net} = \dfrac{d\vec{p}}{dt}$	The most general form of Newton's second law.
Impulse and momentum	$\vec{J} = \vec{F}_{av}\,\Delta t = \Delta\vec{p}$	The relationship between impulse and linear momentum.
The conservation of linear momentum	$\vec{P}_{total,\ before} = \vec{P}_{total,\ after}$	An expression of the conservation of the total linear momentum of a system.
Center of mass	$\vec{R} = \dfrac{m_1\vec{r}_1 + m_2\vec{r}_2 + \cdots + m_N\vec{r}_N}{M}$	The location of the center of mass is the average location of the mass in a system.
Rockets	$v = u_{ex}\ln\left(\dfrac{m_0}{m}\right) - gt$	The speed of a rocket in the presence of gravity.

III. Know Your Units

Quantity	Dimension	SI Unit
Linear momentum ($\vec{p}$)	$\left[MLT^{-1}\right]$	kg·m/s
Impulse ($\vec{J}$)	$\left[MLT^{-1}\right]$	N·s

Practice Problems

1. In the collision shown, the time of impact is from 0 to 28 milliseconds and the force is given by the function: $-6t^2 + 168t$ newtons. To the nearest thousandth of a newton·sec, what is the impulse?

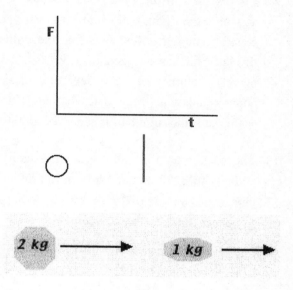

2. The mass of the ball in Problem 1 is 3.74 kg and its velocity before collision is 3.7 m/s. To the nearest hundredth of a m/s, what is its velocity after collision?

3. One lump of clay traveling at 24 m/s overtakes a second lump traveling at 4.2 m/s. After collision they are stuck together. To the nearest tenth of a m/s, what is their common velocity?

4. If the second lump in the above problem was moving to the left before the collision, what would the answer be?

5. One ball traveling at 17.4 m/s strikes a second ball at rest in an elastic, head-on collision. If the mass of the first ball is 27 kg and the mass of the second is 0.5 kg. To the nearest tenth of a m/s, what is the velocity of the first ball after the collision?

6. In the preceding problem, what is the velocity of the second ball after collision?

7. In the figure on the right, the second lowest ball has a mass of 5.2 kg and the next lowest ball has a mass of 6.3 kg. To two decimal places what is X_{CM}?

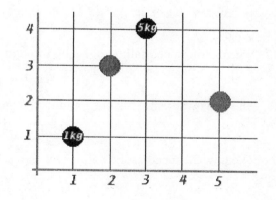

8. An uneven board of length 1.49 meters has a mass density (in kg/m) $\lambda(x) = 4 + 1.31x$. To two decimal places, what is its X_{CM}?

9. The left ball in the figure has a speed of 1.67 m/s and the right ball is at rest prior to an elastic glancing collision. After the collision, the left ball has a speed of 1.36 m/s. To the nearest tenth of a degree, measured counterclockwise from east, what angle does it scatter at if the right ball is scattered at 280°?

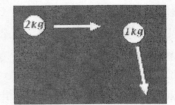

10. The rocket on the right has an exhaust gas velocity of 3122 m/s and a burn rate of 11,371 kg/s. It has a mass of 1.63×10^6 kg. What is its initial acceleration to two decimal places?

Selected Solutions to End of Chapter Problems

5. **(a)** During the firing we use momentum conservation:

$$p_{ri} + p_{bi} = p_{rf} + p_{bf}$$

Because the total momentum is initially zero, we have

$$0 + 0 = (7 \text{ kg})v_{rf} + (10 \times 10^{-3} \text{ kg})(700 \text{ m/s})$$

from which we get $v_{rf} = \boxed{-1.0 \text{ m/s}}$.

(b) The energy transmitted to the shoulder comes from the decrease in kinetic energy of the rifle:

$$\Delta E = -(0 - \tfrac{1}{2} mv^2) = \tfrac{1}{2} (7 \text{ kg})(1.0 \text{ m/s})^2 = \boxed{3.5 \text{ J}}.$$

17. We find the speed falling or rising through a height h from energy conservation:

$$\tfrac{1}{2} mv^2 = mgh \quad \Rightarrow \quad v = \sqrt{2gh}$$

The downward speed as the ball hits the ground is therefore

$$v_1 = \sqrt{2gh_1} = \sqrt{2(9.8 \text{ m/s}^2)(2.0 \text{ m})} = 6.3 \text{ m/s}$$

The upward speed as the ball leaves the ground is

$$v_2 = \sqrt{2gh_2} = \sqrt{2(9.8 \text{ m/s}^2)(1.4 \text{ m})} = 5.2 \text{ m/s}$$

The average force is

$$F_{av} = p/\ t = [m(v_2 - v_1)]/\ t = (0.260 \text{ kg})[(5.2 \text{ m/s}) - (-6.3 \text{ m/s})]/(0.004 \text{ s}) = \boxed{7.5 \times 10^2 \text{ N, up}}.$$

33. We let V be the speed of the block and bullet immediately after the collision and before the pendulum swings. For this perfectly inelastic collision, we use momentum conservation:

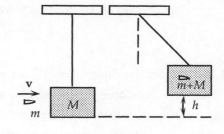

$$mv + 0 = (M + m)V \quad \Rightarrow \quad V/v = m/(M + m).$$

(a) The fractional change in the kinetic energy is

$$\Delta K/K_i = [(M + m)V^2 - mv^2] / mv^2 = [(M + m)/m](V/v)^2 - 1$$

Substituting the earlier result for V/v gives,

$$\Delta K/K_i = -M/(m + M),$$

so the fraction lost = $\boxed{M/(m + M)}$.

(b) For the pendulum motion we use energy conservation:

$$\tfrac{1}{2}(M + m)V^2 = (m + M)gh \quad \Rightarrow \quad V = \sqrt{2gh}$$

We combine this with the result from momentum conservation, $V = mv/(M + m)$, to get

$$V = \sqrt{2gh} = mv/(m + M)$$

which gives

$$\boxed{v = [(m + M)/m]\sqrt{2gh}}.$$

37. (a) Because momentum is conserved, we can write

$$m_1v_1 + m_2v_2 = m_1v_3 + m_2v_4$$

Using the fact that m_2 is initially at rest, we get

$$v_3 = \frac{m_1v_1 - m_2v_4}{m_1} = \frac{(0.4\ \text{kg})(3.0\ \text{m/s}) - (0.8\ \text{kg})(1.6\ \text{m/s})}{0.4\ \text{kg}} = \boxed{-0.2\ \text{m/s}}$$

(b) We determine if kinetic energy is conserved by finding the kinetic energy before and after the collision:

$$K_i = \tfrac{1}{2}m_1v_1^2 = \tfrac{1}{2}(0.4\ \text{kg})(3.0\ \text{m/s})^2 = 1.8\ \text{J}$$

$$K_f = \tfrac{1}{2}m_1v_3^2 + \tfrac{1}{2}m_2v_4^2 = \tfrac{1}{2}(0.4\ \text{kg})(-0.2\ \text{m/s})^2 + \tfrac{1}{2}(0.8\ \text{kg})(1.6\ \text{m/s})^2 = 1.03\ \text{J}.$$

We can see that kinetic energy is lost; therefore, the collision is $\boxed{\text{inelastic}}$.

The maximum kinetic energy is lost when the collision is perfectly inelastic. In that case, we get

$$m_1v_1 + m_2v_2 = (m_1 + m_2)v_f \quad \Rightarrow \quad v_f = \frac{m_1v_1}{m_1 + m_2} = \frac{(0.4\ \text{kg})(3.0\ \text{m/s})}{0.4\ \text{kg} + 0.8\ \text{kg}} = \boxed{1.0\ \text{m/s}}$$

Thus the maximum kinetic energy change is

$$K_{\text{max}} = \tfrac{1}{2}(1.2\ \text{kg})(1.0\ \text{m/s})^2 - 1.80\ \text{J} = -1.20\ \text{J}.$$

The percentage of this that is lost in the actual collision is

$$K/K_{max} = [(1.03 \text{ J} - 1.80 \text{ J})/(-1.20 \text{ J})](100\%) = \boxed{64\%}.$$

61. We ignore the dimensions of the iron block and treat it as a point mass according to the hint given. Since the handle is uniform, we know that its center of mass is at its midpoint. We choose the origin at the bottom of the handle. For the system, we have

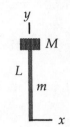

$$X = \Sigma m_i x_i / \Sigma m_i = [(4.0 \text{ kg})(0) + (1.8 \text{ kg})(0)]/(4.0 \text{ kg} + 1.8 \text{ kg}) = \boxed{0}.$$

$$Y = \Sigma m_i y_i / \Sigma m_i = [(4.0 \text{ kg})(1.2 \text{ m}) + (1.8 \text{ kg})(0.6 \text{ m})]/(4.0 \text{ kg kg} + 1.8 \text{ kg})$$

$$= \boxed{1.0 \text{ m from the bottom of the handle}}.$$

75. As described in the textbook, without gravity, we have the equation for the thrust as

$$m\frac{dv}{dt} = -u_{ex}\frac{dm}{dt}$$

In the presence of a constant gravitational force, we will have

$$m\frac{dv}{dt} = -u_{ex}\frac{dm}{dt} - mg$$

Because the gravitational force is assumed constant, when we integrate that term over time, it contributes a term $-gt$ having divided by the mass. The rest remains exactly as worked out in the text; therefore,

$$v = u_{ex} \ln(m_0/m) - gt = u_{ex} \ln[m_0/(m_0 - m_{fuel})] - gt.$$

From the data given, we get

$$v = (2800 \text{ m/s}) \ln[1/(1 - 0.7)] - (9.8 \text{ m/s}^2)(90 \text{ s}) = \boxed{2.5 \times 10^3 \text{ m/s}}.$$

Answers to Practice Quiz

1. (b) **2.** (a) **3.** (e) **4.** (c) **5.** (c) **6.** (a) **7.** (d) **8.** (a) **9.** (e) **10.** (c) **11.** (d) **12.** (c)

Answers to Practice Problems

1. 21.952 N·s   **2.**   −2.17 m/s

3. 17.4 m/s   **4.**   14.6 m/s

5. 16.8 m/s   **6.**   34.2 m/s

7. 3.31   **8.**   0.79 m

9. 29.8°   **10.** 11.98 m/s^2

CHAPTER 9

ROTATIONS OF RIGID BODIES

Chapter Objectives

After studying this chapter, you should

1. be able to describe the rotational motion of a rigid body in terms of its angular velocity and angular acceleration vectors.
2. be able to calculate the rotational inertia of discreet and continuous distributions of matter.
3. be able to calculate the rotational kinetic energy of bodies in pure rotation and rolling objects.
4. be able to determine the torque about a particular axis due to the action of a force.
5. be able to apply Newton's second law for rotation for the analysis of rotational motion.
6. be able to understand the concept of angular momentum.

Chapter Review

In this chapter, and the next, we study rotational kinematics and dynamics. Rotational motion is every bit as important as the linear (or translational) motion that we've been studying thus far. As you read these chapters, try to notice how the study of rotational motion parallels that of translational motion. Many of the concepts as well as the mathematical treatments are the same, we need only to alter the physical interpretation of these concepts from that of a translating body to a rotating one.

9–1 Simple Rotations of a Rigid Body

A rigid body is a body within which the relative positions of the points in the body are fixed. Here we consider such a body rotating about a single fixed axis passing through the body. This motion is more easily described using plane polar coordinates (see section 3–5). Each point in the body remains at a fixed distance r from the axis and the rotation of these points can be described by a single angular variable $\theta(t)$ measured in radians.

The **angular velocity** of the body is obtained in a way similar to how linear velocity is obtained. The magnitude of the average angular velocity over a time interval Δt is

$$\omega_{av} = \frac{\Delta \theta}{\Delta t}.$$

The SI unit of angular velocity is rad/s = s^{-1} (since radian is dimensionless). The instantaneous angular velocity is

$$\vec{\omega} = \frac{d\theta}{dt} \hat{\omega},$$

where $\hat{\omega}$ is a unit vector in the direction of the angular velocity. The direction of the angular velocity is along the axis of rotation. Which way it points along the axis can be determined by a *right-hand rule*:

> *Curl the fingers of your right hand around the axis along the direction of the rotation of the body; then, your extended thumb indicates which way along the axis the angular velocity points.*

It is common to define the axis about which the object rotates as the *z*-axis. If this choice is adopted, then the angular velocity points in the positive *z* direction for counterclockwise rotation (as viewed from above the *x-y* plane) and along the negative *z* direction for clockwise rotation.

The **angular acceleration** of the body is also defined in a way similar to the definition of translational acceleration. The average acceleration is

$$\vec{\alpha}_{av} = \frac{\Delta\vec{\omega}}{\Delta t} \, ;$$

and the instantaneous angular acceleration is

$$\vec{\alpha} = \frac{d\vec{\omega}}{dt} = \frac{d^2\theta}{dt^2}\hat{\alpha} \, ,$$

where $\hat{\alpha}$ is a unit vector in the direction of the angular acceleration. This direction is determined by the rate of change of $\vec{\omega}$. For the present case of rotation about a fixed axis passing through the rigid body, if the angular acceleration is such that ω is increasing, then $\hat{\alpha} = \hat{\omega}$; if the angular acceleration is such that ω is decreasing, $\hat{\alpha} = -\hat{\omega}$. The SI unit of angular acceleration is rad/s^2.

One thing you may have noticed about the three quantities used to described this motion (θ, ω, α), is that the mathematical relationships among them are exactly the same as the relationships among the corresponding quantities for translational motion (x, v, a). This means that we can follow the prescription already laid out for translational motion. All the equations used previously for describing motion with constant velocity and constant acceleration in one dimension also apply to the rotational motion of a rigid body about a fixed axis. The only mathematical difference is that the names of the variables are changed in the following way:

$$x \to \theta, \;\; v \to \omega, \;\; a \to \alpha, \;\; t \to t$$

Of course, along with changing the variables, you should also change the physical picture in your mind of the type of motion you are describing. Nevertheless, the mathematical descriptions are identical.

With the above comments in mind, we consider the special case of motion at a constant angular velocity $\omega(t) = \omega_0$, we have the familiar relationship that the change in position (this time angular position) depends linearly on time

$$\theta - \theta_0 = \omega_0 t \, .$$

The period of the motion, which is the amount of time for one full rotation, is given by

$$T = \frac{2\pi}{\omega_0} \, ,$$

and the frequency is

$$f = \frac{1}{T} = \frac{\omega_0}{2\pi} \, .$$

For the special case of motion with constant angular acceleration, we can see that the same replacements produce four equations just like the ones we used for straight-line motion in Chapter 2,

$$\omega = \omega_0 + \alpha t$$

$$\omega_{av} = \tfrac{1}{2}(\omega_0 + \omega)t$$

$$\theta = \theta_0 + \omega_0 t + \tfrac{1}{2}\alpha t^2$$

$$\omega^2 = \omega_0^2 + 2\alpha(\theta - \theta_0)$$

A point on such a rotating body has both a radial (i.e., centripetal) acceleration and, if $\alpha \neq 0$, a tangential acceleration. The radial acceleration is given by

$$a_{radial} = \frac{v^2}{r},$$

and the tangential acceleration is

$$a_{tan} = \alpha r .$$

Example 9–1 On a small fan in your room, the tip on one of the blades of the fan can go through 1.25 revolutions in one second. Suppose the blades of this fan rotate with constant angular acceleration for the first 1.00 seconds. **(a)** What is the instantaneous angular velocity of the blades at $t = 1.00$ seconds after being turned on, and **(b)** what is the angular acceleration?

Setting it up: The information given in the problem is the following
Given: $\theta = 1.25$ rev , $t = 1.00$ s;   **Find**: **(a)** ω, **(b)** α

Strategy:
In both parts we make use of the expressions for uniformly accelerated motion to obtain a solution. Let's make use of our freedom to define $\theta_0 = 0$ and $t_0 = 0$. Also, notice that since the fan is just being turned on, $\omega_0 = 0$.

Working it out:
(a) Based on the given information, an appropriate expression for calculating ω is

$$\theta = \theta_0 + \tfrac{1}{2}(\omega_0 + \omega)t \;\Rightarrow\; \theta = \tfrac{1}{2}\omega t .$$

Therefore,

$$\omega = \frac{2\theta}{t} = \frac{2(1.25\,\text{rev})(2\pi\,\text{rad/rev})}{1.00\,\text{s}} = 15.7\ \text{rad/s} .$$

(b) Based on the given information, an appropriate expression for calculating α is

$$\omega = \omega_0 + \alpha t \;\Rightarrow\; \alpha = \frac{\omega}{t}$$

Therefore,

$$\alpha = \frac{15.71 \text{ rad/s}}{1.00 \text{ s}} = 15.7 \text{ rad/s}^2 .$$

What do you think? If the blades are 7.00 cm long, what is the acceleration of the tip of a blade?

Practice Quiz

1. Which quantity is used to describe how rapidly an object rotates?
    (a) angular position   **(b)** angular velocity   **(c)** angular acceleration   **(d)** none of the above

2. If an object is seen to rotate faster and faster, which quantity best describes this aspect of its motion?
    (a) angular position   **(b)** angular velocity   **(c)** angular acceleration   **(d)** none of the above

3. How long would it take an object rotating at a constant speed of 17.0 rad/s to rotate through 235 radians?
    (a) 0.0723 s   **(b)** 3.4 N   **(c)** 5.0 N   **(d)** 13.8 s   **(e)** 3.1 N

4. If an object rotates at 2.50 rpm, what is the equivalent angular speed in rad/s?
    (a) 2.50   **(b)** 15.7   **(c)** 0.262   **(d)** 0.0417   **(e)** 25.0

9–2—9–3 Rotational Kinetic Energy and Rotational Inertia

When studying translational motion, we introduced the idea of kinetic energy, which is an energy associated with "mass in motion." Objects that are rotating but not translating also have mass in motion. We should, therefore, be able to write its kinetic energy in terms of the rotational quantities. In doing this, we find that the rotational kinetic energy of a system is

$$K_{rot} = \tfrac{1}{2}\left(\sum m_i r_i^2\right)\omega^2$$

Comparison between this expression and the translational form, $K = \tfrac{1}{2}mv^2$, suggests that the quantity $\sum mr^2$ plays a role similar to mass. We call this quantity the **rotational inertia,** I of the system (also called *moment of inertia*):

$$I = \sum m_i r_i^2$$

The SI unit of the moment of inertia is kg·m^2. Therefore, the rotational kinetic energy can be written as

$$K = \tfrac{1}{2}I\omega^2 .$$

The above expression for the rotational inertia is convenient for calculating I when the number of particles in the system is small. However, for systems with so many particles that we should treat the system as a continuous distribution of matter, we perform an integration instead of a summation. For some of the more commonly encountered objects, values of rotational inertia are provided in Table 9–1 of your textbook (pg. 253). Notice that I depends on the axis about which the object is rotating, so that the same object can have many different values of I associated with it. You should become very familiar with Table 9-1.

When evaluating the rotational inertia for continuous distributions of matter we replace the mass of the particle by a differential mass element dm written in terms of a mass density in the same way as it is done when calculating the center of mass of continuous distributions.

$$dm = \lambda dx \quad \text{for linear distributions}$$
$$= \sigma dA \quad \text{for two-dimensional distributions}$$
$$= \rho dV \quad \text{for three-dimensional distributions.}$$

The rotational inertia is then calculated as

$$I = \int R^2 dm \,,$$

where R^2 is the distance from the axis about which I is being calculated.

Once the rotational inertia of an object about an axis through its center of mass is known, we can easily find the rotational inertia about any other axis parallel to it using the parallel-axis theorem

$$I_{pa} = I_{cm} + Md^2 \,.$$

In this expression, I_{cm} and I_{pa} are the rotational inertia about axes through the center of mass and a parallel axis, respectively, M is the mass of the object and d is the distance between the axes.

Example 9–2 A solid cylinder, of mass 2.5 kg and radius 36 cm, rotates about an axis through its surface with an angular velocity of 3.3 rad/s. Determine its rotational kinetic energy.

Setting it up: A diagram is helpful here. Notice that the axis is parallel to an axis passing through the center-of-mass. The information given in the problem is the following:

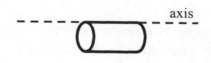

Given: $m = 2.5$ kg, $R = 36$ cm, $\omega = 3.3$ rad/s

Find: K

Strategy:
To calculate the rotational kinetic energy, we need the values of I and ω. Since the value of ω is given the key to the solution is determining the correct rotational inertia. This can be done using the parallel-axis theorem.

Working it out:
From Table 9–1 in your textbook we see that for a solid cylinder $I_{cm} = MR^2/2$. Applying the parallel-axis theorem, the rotational inertia about an axis through the surface is

$$I = \tfrac{1}{2}MR^2 + MR^2 = \tfrac{3}{2}MR^2 \,.$$

Therefore, the rotational kinetic energy is

$$K = \tfrac{1}{2}I\omega^2 = \tfrac{3}{4}MR^2\omega^2 = \tfrac{3}{4}(2.5 \text{ kg})(0.36 \text{ m})^2(3.3 \text{ rad/s})^2 = 2.6 \text{ J} \,.$$

Practice Quiz

5. Two solid wheels have the same mass, but one has twice the radius of the other. If they both rotate about axes through their centers, the one with the larger radius will have

(a) half the rotational inertia of the other one

(b) the same rotational inertia as the other one

(c) twice the rotational inertia of the other one

(d) four times the rotational inertia of the other one

(e) eight times the rotational inertia of the other one

6. A thin rod of mass 1.0 kg and length 1.0 m rotates about an axis through its center at 1.0 revolutions per second. What is its kinetic energy?

(a) 1.6 J   (b) 1.0 J   (c) 0.083 J   (d) 39 J   (e) 20 J

9–4 Torque

If you want to cause a stationary object to move, you must apply a force. Similarly, if you want to cause a nonrotating object to rotate, you must also apply a force. However, in the rotational case, not just any force will work. Crucial to determining whether rotation will result from a force are two factors: **(a)** how the applied force is directed relative to the axis of rotation and **(b)** the distance of the point at which the force is being applied from the axis. These two factors combine with the magnitude of the force to form a quantity called **torque**, $\vec{\tau}$. It is through the application of a torque that an object will begin to rotate. The magnitude of the torque τ that results from the application of a force $\vec{F}$ is given by

$$\tau = rF\sin\theta ,$$

where r is the magnitude of a radial vector $\vec{r}$ from the axis of rotation to the point of action of the force, and θ is the smallest angle between $\vec{r}$ and $\vec{F}$. The SI unit of torque is the N·m. *Note that in calculations dealing with torque the N·m is **not** called a joule.*

For convenience, the expression for torque is often looked at in two ways. The quantity $F\sin\theta$ equals the component of $\vec{F}$ tangential to a circle of radius r centered on the axis of rotation, F_t, so we can write $\tau = rF_t$. Grouped the other way, the quantity $r\sin\theta$ equals the component of $\vec{r}$ perpendicular to the line of force, $r_\perp$, called the **lever arm** (or *moment arm*) of the force, so we can also write $\tau = r_\perp F$.

Torque is the quantity that plays the role of force for rotational motion. Just as force causes translational acceleration, torque causes angular acceleration. The relationship between $\vec{\tau}$ and $\vec{\alpha}$ is very similar to that between force and translational acceleration:

$$\vec{\tau}_{net} = I\vec{\alpha} .$$

We can see how strong the similarity is by recalling that the rotational inertia, I, is the rotational quantity that acts like mass does for translational motion. The above equation is sometimes referred to as Newton's second law for rotation.

It is clear from the above equation that the direction of the net torque is the same as that of the angular acceleration. The direction of the torque from an individual force can be determined from two facts; (a) $\vec{\tau}$

points along the line perpendicular to the plane formed by $\vec{r}$ and $\vec{F}$; (b) the direction along this line is determined using a *right-hand rule*:

> *Align the fingers of your right hand along $\vec{r}$ such that your fingers can curl toward $\vec{F}$ through θ; your extended thumb then indicates the direction of the torque along the perpendicular line.*

The analysis of situations involving extended bodies proceeds much in the same way as for particles. The key difference is that because extended bodies can rotate depending on where and how the forces are applied, we must account for this fact in our analysis. Therefore, a proper free-body diagram of an extended object will also indicate the location at which the forces act on the body. Having done this, a convenient axis must be chosen about which to calculate the torques resulting from these forces. Then, Newton's second law for both translation and rotation can be applied to determine the resulting behavior of the body.

When it comes to indicating where on a body a force acts, gravity creates a special circumstance because the gravitational force acts upon every part of the body simultaneously. This case presents another example of the usefulness of the center-of-mass concept. As discussed in Chapter 8, there are situations in which a system of particles can be treated as a single particle with all of its mass located at its center of mass. Concerning the effect of gravity on extended bodies, we have two such examples here: **(a)** The torque about any axis due to the weight of an extended body equals the torque that would result from a particle of equal mass located at the center-of-mass of the body; **(b)** The change in the gravitational potential energy of an extended body equals the change that would result if the body were a single particle with all of its mass located at the center-of-mass. Therefore, if a body of mass M moves such that its center-of-mass undergoes a change in height Δh_{cm}, then the change in gravitational potential energy of the body is $\Delta U = Mg(\Delta h_{cm})$.

Example 9–3 A decorative ornament is hung from the end of an 82.3-cm-long horizontal beam attached to a wall as shown below. The beam weighs 12.1 N, and the ornament weighs 7.62 N. If the cord is not strong enough to support the structure and breaks, what is the initial angular acceleration of the beam? (Assume the ornament remains attached to the end of the beam).

Setting it up:

The diagram is shown at the right. The cord attached to the beam breaks. The following information is given in the problem:

Given: $L = 82.3$ cm, $W_b = 12.1$ N, $W_o = 7.62$ N

Find: α

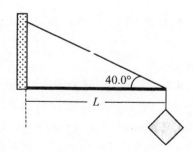

Strategy:

Because the cord breaks, the net torque is not equal to zero in this case, so we use Newton's second law for rotation, $\vec{\tau}_{net} = I\vec{\alpha}$.

Working it out:

The moment of inertia, calculated about an axis through the end of the beam, is given by both the beam and the ornament: $I = I_b + I_o$. For the beam, we use the results in Table 9–1

$$I_b = \tfrac{1}{3} M_b L^2$$

and for the ornament we have just $I_o = M_o L^2$. Therefore,

$$I = \left(M_o + \frac{1}{3} M_b \right) L^2$$

The net torque about this same axis is due to the weight of the beam and the weight of the ornament.

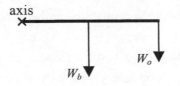

From the right-hand rule for the direction of a torque, we can see that the lever arm for each force points horizontally to the right with $r_b = L/2$ and $r_o = L$. Therefore, each torque points into the page; let's call that the negative z direction. So, the net torque is

$$\vec{\tau}_{net} = \sum \vec{\tau}_i = -W_b \tfrac{L}{2} \hat{k} - W_o L \hat{k}.$$

Combining this with Newton's second law for rotation, we have

$$-L \left(W_o + W_b / 2 \right) \hat{k} = L^2 \left(M_o + M_b / 3 \right) \vec{\alpha}.$$

Solving this for the angular acceleration gives

$$\vec{\alpha} = -\frac{W_o + W_b / 2}{\left(M_o + M_b / 3 \right) L} \hat{k}.$$

Making use of the fact that $M = W/g$, we can solve for the angular acceleration,

$$\vec{\alpha} = -\hat{k} \frac{7.62 \text{ N} + (12.1 \text{ N})/2}{\left[7.62 \text{ N} + (12.1 \text{ N})/3 \right](0.823 \text{ m})/9.80 \text{ m/s}^2} = -(14.0 \text{ rad/s}^2) \hat{k}.$$

What do you think?   Would this value of the angular acceleration remain the same as the beam rotates?

Practice Quiz

7.   In trial 1, a force $\vec{F}$ is applied tangentially to the rim of a wheel, causing it to rotate about its center. In trial 2, an equal force is applied tangentially at a point halfway from the center to the rim, also causing a rotation about the center. How do the torques applied in each trial relate to each other?

    (a) Trial 2 has twice the torque of trial 1.

    (b) Trial 1 has twice the torque of trial 2.

    (c) Trial 1 has half the torque of trial 2.

    (d) The torques are equal.

8. If an object has a constant net torque applied to it, this implies that the object must also have
   (a) constant moment of inertia
   (b) constant angular velocity
   (c) constant angular acceleration
   (d) constant angular displacement
   (e) none of the above

9. A uniform hoop of radius 0.75 m and mass 2.25 kg rotates with an angular acceleration of 44 rad/s^2 about an axis through its center. What is the net torque on this hoop?
   (a) 0 N·m (b) 33 N·m (c) 1.3 N·m (d) 74 N·m (e) 56 N·m

9–5 Angular Momentum and Its Conservation

In studying translational motion we found that the concept of linear momentum was very useful and important for understanding the behavior of objects, especially systems of particles. For similar reasons, we also have a rotational form of this concept called **angular momentum**, $\vec{L}$. Just as for linear momentum, there is a conservation principle for angular momentum. The angular momentum of a symmetrical object that is rotating about its axis of symmetry with an angular velocity $\vec{\omega}$ is given by

$$\vec{L} = I\vec{\omega}.$$

This result is a direct analogue to linear momentum $\vec{p} = m\vec{v}$. As can be seen from the above equation, the SI unit of angular momentum is kg·m^2/s, which is equivalent to a joule-second (J·s).

The parallel role angular momentum and linear momentum can be strengthened by noting that Newton's second law for rotation, takes its most complete form in terms of angular momentum as

$$\vec{\tau}_{net} = \frac{d\vec{L}}{dt},$$

which is comparable to the translational equation $\vec{F} = d\vec{p}/dt$. Furthermore, the rotational kinetic energy of this symmetrical object can be written in terms of its angular momentum

$$K_{rot} = \frac{L^2}{2I},$$

which is comparable to the translational equation $K_{trans} = p^2/2m$. The value of the concept of angular momentum, including its conservation law, will be explored in more detail in Chapter 10.

9–6 Rolling

Rolling motion represents an important application that combines both rotational and translational motion. For a wheel that is free to roll, without slipping, each point on the wheel rotates about the wheel's axle while also being carried forward from the overall translation of the wheel. As a result of this combination of rotational and translational speeds, the point of contact between the wheel and the ground is instantaneously at rest (otherwise there would be slipping), the center of the wheel moves forward at

$v_{cm} = R\omega$ (R – radius, ω – angular velocity), and the point at the top of the wheel moves at twice this speed: $v_{top} = 2R\omega$.

The kinetic energy of the wheel equals the sum of its rotational kinetic energy about the center-of-mass and the translational kinetic energy of the wheel as a whole. The translational kinetic energy of the wheel as a whole is found by treating the wheel as a point with all of its mass located at its center-of-mass. Therefore, we have

$$K = \tfrac{1}{2}I_{cm}\omega^2 + \tfrac{1}{2}Mv_{cm}^2 .$$

For a symmetrical rolling wheel or cylinder, the dynamics of the motion is governed by applying the translational form of Newton's second law to the center-of-mass

$$\vec{F}_{net} = M\vec{a}_{cm} ,$$

and the rotational form of Newton's second law about the center-of-mass,

$$\vec{\tau}_{cm} = I_{cm}\vec{\alpha} .$$

If the object is rolling without slipping, then the frictional force is static friction at the point (or line) of contact between the surfaces. Generally, gravity and the normal force will not contribute to the net torque about the center-of-mass.

Example 9–5 A solid ball of mass 1.40 kg and radius 0.391 m is released from rest and caused (by gravity) to roll from the top of a track 10.0 m high down to a height of 0.500 m above the ground, where it is launched vertically upward. How high in the air will the ball go assuming it continues to spin after leaving the track (neglect any friction)?

Setting it up:
A diagram is helpful. The ball is at rest at the top and is launched vertically upward. The information given in the problem is as follows:
Given: $m = 1.40$ kg, $r = 0.391$ m, $v_0 = 0$, $v_f = 0$,
     $H = 10.0$ m, $h = 0.500$ m;
Find: y_{max}

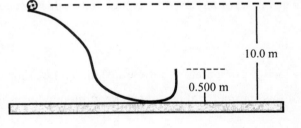

Strategy:
Because the ball rolls without slipping, we can apply mechanical energy conservation. Let's take the bottom of the track (the ground) as the reference level for gravitational potential energy.

Working it out:
The mechanical energy before the ball is released is in the form of only gravitational potential energy

$$E = mgH .$$

The final energy, at height y_{max}, is a combination of potential and rotational kinetic energy

$$E = mgy_{max} + \tfrac{1}{2}I_{cm}\omega^2 .$$

To apply this result, we need to determine ω when the ball leaves the track. The energy when it leaves is

$$E = mgh + \tfrac{1}{2}mv_{cm}^2 + \tfrac{1}{2}I_{cm}\omega^2 .$$

From Table 9–1 in the textbook, I_{cm} is given by $I_{cm} = \tfrac{2}{5}mr^2$. We further use the fact that $v_{cm} = r\omega$. We can substitute these results into the equation for the energy when the ball leaves the track to get

$$E = mgh + \tfrac{1}{2}mr^2\omega^2 + \tfrac{1}{5}mr^2\omega^2 = mgh + \tfrac{7}{10}mr^2\omega^2 .$$

Setting this equal to the initial total energy gives

$$mgH = mgh + \tfrac{7}{10}mr^2\omega^2 \ \Rightarrow\ \omega = \left[\frac{g(H-h)}{\tfrac{7}{10}r^2}\right]^{1/2} = \left[\frac{(9.80 \text{ m/s}^2)(9.50 \text{ m})}{\tfrac{7}{10}(0.391 \text{ m})^2}\right]^{1/2} = 29.50 \text{ rad/s} .$$

With this result, we can now go back to the equation for the energy when it is at its maximum height in the air and set it equal to the initial total energy

$$mgH = mgy_{max} + \tfrac{1}{5}mr^2\omega^2$$

$$\therefore y_{max} = H - \frac{r^2\omega^2}{5g} = 10.0 \text{ m} - \frac{(0.391 \text{ m})^2 (29.50 \text{ rad/s})^2}{5(9.80 \text{ m/s}^2)} = 7.28 \text{ m}.$$

What do you think?   What is the linear speed of the ball when it leaves the track?

Practice Quiz

10. If a bicycle moves at 6.5 m/s and its wheels have a diameter of 70 cm, how fast are the wheels rotating?

    (a) 9.3 rad/s   (b) 19 rad/s   (c) 4.5 rad/s   (d) 2.3 rad/s   (e) 0.093 rad/s

11. A hollow sphere of mass 0.22 kg and radius 0.16 m rolls along the ground with a constant linear speed of 2.0 m/s. What is its kinetic energy?

    (a) 0.0075 J   (b) 0.29 J   (c) 0.070 J   (d) 0.73 J   (e) 0.44 J

Reference Tools and Resources

I. Key Terms and Phrases

angular velocity the rate of change of angular position

angular acceleration the rate of change of angular velocity

rotational inertia a quantity that represents the inertial property of a rotating object or system

torque the combination of a force and how it is applied that causes angular acceleration

lever arm the perpendicular distance from the axis of rotation to the line of force used to calculate torque

angular momentum a vector quantity that represents a rotational analog of linear momentum

II. Important Equations

Name/Topic	Equation	Explanation
Angular velocity	$\vec{\omega} = \dfrac{d\theta}{dt}\hat{\omega}$	The definition of angular velocity.
Angular acceleration	$\vec{\alpha} = \dfrac{d\vec{\omega}}{dt}$	The definition of angular acceleration.
Constant angular velocity	$\theta - \theta_0 = \omega_0 t$	The equation for motion with constant angular velocity.
Constant angular acceleration	$\omega = \omega_0 + \alpha t$ $\omega_{av} = \frac{1}{2}(\omega_0 + \omega)t$ $\theta = \theta_0 + \omega_0 t + \frac{1}{2}\alpha t^2$ $\omega^2 = \omega_0^2 + 2\alpha(\theta - \theta_0)$	A system of equations to describe motion with constant angular acceleration.
Rotational kinetic energy	$K = \frac{1}{2}I\omega^2$	The kinetic energy of a rotating system.
Rotational inertia	$I = \sum m_i r_i^2$ or $I = \int R^2 dm$	The rotational inertias of discreet and continuous distributions of matter, respectively.
Torque	$\tau = rF\sin\theta$	The magnitude of the torque due to a force.
Torque	$\vec{\tau}_{net} = I\vec{\alpha}$	The relationship between torque and angular acceleration.
Angular momentum	$\vec{L} = I\vec{\omega}$	The angular momentum of a rigid body rotating about a fixed axis.
Angular momentum	$\vec{\tau}_{net} = \dfrac{d\vec{L}}{dt}$	The relationship between torque and angular momentum.
Rolling	$v_{cm} = R\omega$	The speed of the center of mass of a rolling object.
Rolling	$K = \frac{1}{2}I_{cm}\omega^2 + \frac{1}{2}Mv_{cm}^2$	The kinetic energy of a rolling object is the sum of its rotational and translational kinetic energies.

III. Know Your Units

Quantity	Dimension	SI Unit
Angular position (θ)	dimensionless	rad
Angular velocity ($\vec{\omega}$)	$\left[T^{-1}\right]$	rad/s
Angular acceleration ($\vec{\alpha}$)	$\left[T^{-2}\right]$	rad/s^2
Rotational inertia (I)	$\left[ML^2\right]$	kg·m^2
Torque ($\vec{\tau}$)	$\left[ML^2T^{-2}\right]$	N·m
Angular momentum ($\vec{L}$)	$\left[ML^2T^{-1}\right]$	kg·m^2/s

Practice Problems

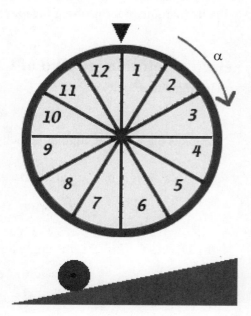

1. The wheel of fortune is divided into twelve 30° (0.5236 radians) sections. When the wheel is spun counter-clockwise it will have a negative angular acceleration of 0.608 rad/s^2. The jackpot is in position 9. To the nearest hundredth of a rad/s, what is the minimum angular velocity you need to give it in order to win the jackpot?

2. Now you are eligible to try for the grand prize. It is located at 6. The catch is, you must go by it one full revolution before stopping at it. What is the minimum angular velocity needed?

3. A solid disk rolls up a hill. It is revolving at 76 rpm when it starts up the hill. To two decimal places in radians/s, what is its angular velocity?

4. The disk in the previous problem has a negative angular acceleration of 0.85 rad/s^2. To the nearest hundredth of a second, how long will it take to stop?

5. The radius of the disk in Problem 4 is 0.4 meters. To three decimal places, what is the magnitude of its linear acceleration to three decimal places?

6. Using conservation of energy, what is the final height of the disk, to the nearest hundredth of a meter?

7. To the nearest hundredth of a meter, how far up the hill does the disk travel before stopping?

8. To the nearest thousandth of a degree, what is the angle of the incline in Problem 3?

9. A 1-kg mass hangs by a string from a disk with radius 10.4 cm which has a rotational inertia of 5×10^{-5} kg·m². After it falls a distance of 0.5 meters, how fast is it going to the nearest hundredth of a m/s?

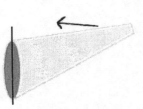

10. A tapered pole, shown above, of length 3.27 meters has a mass density (in kg/m) of $\lambda(x) = 4.7 - 0.27x$. To two decimal places, what is its rotational inertia about the axis shown?

Selected Solutions to End of Chapter Problems

9. The angular velocity after 5 s is

$$\omega_1 = \omega_0 + \alpha_1 t = 0 + \left(0.4 \text{ rad/s}^2\right)(5 \text{ s}) = 2.0 \text{ rad/s}$$

which is the constant angular velocity for the next 30 s.

(a) Because it is moving at a constant angular velocity of 2.0 rad/s, after 20 s, we have

$$\alpha_{av} = \Delta\omega / \Delta t = (2.0 \text{ rad/s} - 0)/(20 \text{ s}) = \boxed{0.10 \text{ rad/s}^2}$$

(b) During the first 5 s, the angle turned through is

$$\theta_1 = \omega_0 t + \tfrac{1}{2}\alpha_1 t^2 = 0 + \tfrac{1}{2}(0.4 \text{ rad/s}^2)(5 \text{ s})^2 = 5.0 \text{ rad}$$

During the next 30 s, the angle turned through is

$$\theta_2 = \omega_1 t_2 = (2.0 \text{ rad/s})(30 \text{ s}) = 60 \text{ rad}$$

Because the slowing down is the reverse of the initial motion, the angle turned through is

$$\theta_3 = 5.0 \text{ rad}$$

The total number of revolutions is

$$\theta_{total} = \theta_1 + \theta_2 + \theta_3 = (5 \text{ rad} + 60 \text{ rad} + 5 \text{ rad})/(2\pi \text{ rad/rev}) = \boxed{11 \text{ rev}}$$

(c) The child will travel

$$s = r\theta = (3 \text{ m})(70 \text{ rad}) = \boxed{2.1 \times 10^2 \text{ m}}$$

23. From the diagram, we see that the distance of the ball from the axis is $r = L\sin\theta$. The rotational inertia of the ball about the axis is

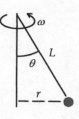

$$I = mr^2 = m(L\sin\theta)^2$$

The rotational kinetic energy is then

$$K = \tfrac{1}{2}I\omega^2 = \tfrac{1}{2}mL^2\omega^2\sin^2\theta = \tfrac{1}{2}(0.75\text{ kg})(1.5\text{ m})^2\left(25\tfrac{\text{rad}}{\text{s}}\right)^2\sin^2 30° = \boxed{1.3\times10^2\text{ J}}$$

At an angle of 60°, we have

$$K = \tfrac{1}{2}I\omega^2 = \tfrac{1}{2}mL^2\omega^2\sin^2\theta = \tfrac{1}{2}(0.75\text{ kg})(1.5\text{ m})^2\left(25\tfrac{\text{rad}}{\text{s}}\right)^2\sin^2 60° = \boxed{4.0\times10^2\text{ J}}$$

29. The density of the thick cylinder is $\rho = M/[\pi L(R_2^2 - R_1^2)]$. Let I_{tot} be the rotational inertia of a completely solid cylinder having the same density, I_1 the rotational inertia of the inner part (of radius R_1), and I be the sought after rotational inertia of the cylinder. If it were solid, then $I_{\text{tot}} = I_1 + I$. So that $I = I_{\text{tot}} - I_1$. First writing mass as density times volume, the rotational inertias of the solid and inner cylinders would be

$$I_{\text{tot}} = \tfrac{1}{2}MR_2^2 = \tfrac{1}{2}\left(\rho\pi R_2^2 L\right)R_2^2 = \tfrac{1}{2}\rho\pi L R_2^4$$

$$I_1 = \tfrac{1}{2}MR_1^2 = \tfrac{1}{2}\left(\rho\pi R_1^2 L\right)R_1^2 = \tfrac{1}{2}\rho\pi L R_1^4$$

Therefore,

$$I = I_{\text{tot}} - I_1 = \tfrac{1}{2}\rho\pi L R_2^4 - \tfrac{1}{2}\rho\pi L R_1^4 = \tfrac{1}{2}\rho\pi L\left(R_2^4 - R_1^4\right)$$

If we now substitute in the expression for the density we get

$$I = \tfrac{1}{2}\rho\pi L\left(R_2^4 - R_1^4\right) = \frac{\pi L}{2}\left[\frac{M}{\pi L\left(R_2^2 - R_1^2\right)}\right]\left(R_2^4 - R_1^4\right) = \frac{M}{2}\frac{\left(R_2^2 + R_1^2\right)\left(R_2^2 - R_1^2\right)}{\left(R_2^2 - R_1^2\right)} = \boxed{\tfrac{1}{2}M\left(R_2^2 + R_1^2\right)}.$$

47. From the angular motion of the wheel we find the angular acceleration,

$$\omega = \omega_0 + \alpha t \implies \alpha = (\omega - \omega_0)/t = (4.15\text{ rad/s} - 0)/(2.68\text{ s}) = \boxed{1.55\tfrac{\text{rad}}{\text{s}^2}}$$

(a) We write $\Sigma\tau = I\alpha$ about the axis:

$$FR = I\alpha \implies I = FR/\alpha = (13.4\text{ N})(0.246\text{ m})/\left(1.55\tfrac{\text{rad}}{\text{s}^2}\right) = \boxed{2.13\text{ kg}\cdot\text{m}^2}$$

(b) $\Delta L = I(\Delta\omega) = (2.13\text{ kg}\cdot\text{m}^2)(4.15\text{ rad/s} - 0) = \boxed{8.84\tfrac{\text{kg}\cdot\text{m}^2}{\text{s}}}$

(c) $\theta = \omega_0 t + \tfrac{1}{2}\alpha t^2 = 0 + \tfrac{1}{2}\left(1.55\tfrac{\text{rad}}{\text{s}^2}\right)(2.68\text{ s})^2 = 5.57\text{ rad} = \boxed{0.886\text{ rev}}$

(d) $K = \tfrac{1}{2}I\omega^2 = \tfrac{1}{2}(2.13\text{ kg}\cdot\text{m}^2)(4.15\text{ rad/s})^2 = \boxed{18.3\text{ J}}$

49. We choose the initial direction of the wheel's angular velocity, upward, as positive. Therefore, its final angular velocity is negative. The angular momentum in the direction of the axis is conserved:

$$I_w\omega_1 + 0 = I_w\omega_2 + I_s\omega_s \quad \Rightarrow \quad \omega_s = \frac{I_w(\omega_1 - \omega_2)}{I_s} = \frac{(1.2\,\text{kg}\cdot\text{m}^2)[4\pi - (-4\pi)]}{(8\,\text{kg}\cdot\text{m}^2)} = \boxed{3.8\,\text{rad/s up}}$$

Thus the student will $\boxed{\text{rotate in the direction of the original rotation of the wheel}}$.

57. We write $\Sigma F_x = ma_x$ from the force diagram for the cylinder:

$$Mg\sin\theta - f = Ma$$

Both the normal force and the weight pass through the center of mass of the wheel and, therefore, apply no torque about it. Writing $\Sigma\tau = I\alpha$ about the center of mass:

$$fR = I\alpha.$$

Noting that for the rolling motion $a = R\alpha$, we can successively eliminate α and f, to get

$$a = \frac{g\sin\theta}{1 + I/MR^2}$$

For a cylindrical shell, $I = MR^2$; so

$$a = \tfrac{1}{2}g\sin\theta.$$

For the linear motion we use

$$x = v_0 t + \tfrac{1}{2}at^2 = 0 + \tfrac{1}{2}\left[\tfrac{1}{2}(9.8\,\text{m/s}^2)\sin 20°\right](4\,\text{s})^2 = \boxed{13.4\,\text{m}}$$

For the solid cylinder, $I = \tfrac{1}{2}MR^2$; so $a = \tfrac{2}{3}g\sin\theta$.

For the linear motion we use

$$x = v_0 t + \tfrac{1}{2}at^2 = 0 + \tfrac{1}{2}\left[\tfrac{2}{3}(9.8\,\text{m/s}^2)\sin 20°\right](4\,\text{s})^2 = \boxed{17.9\,\text{m}}$$

Answers to Practice Quiz

1. (b) **2.** (c) **3.** (d) **4.** (c) **5.** (d) **6.** (a) **7.** (b) **8.** (c) **9.** (e) **10.** (b) **11.** (d)

Answers to Practice Problems

1.	2.26 rad/s	**2.**	3.29 rad/s
3.	7.96 rad/s	**4.**	9.36 s
5.	0.340 m/s^2	**6.**	0.78 m
7.	14.90 m	**8.**	2.981°
9.	3.12 m/s	**10.**	47.06 kg·m^2

CHAPTER 10

MORE ON ANGULAR MOMENTUM AND TORQUE

Chapter Objectives

After studying this chapter, you should

1. be able to calculate angular momentum and torque using the vector product.
2. be able to analyze the dynamics of rotational motion about points other than the center of mass.
3. be able to use the angular impulse-momentum theorem and apply the conservation of angular momentum.
4. be able to apply the concepts of work and energy to rotational motion.
5. be able to understand the precession of rotating bodies. (optional)

Chapter Review

In this chapter, we continue the study of rotation by looking at the dynamics of rotational motion. It is here that we introduce the most general forms of angular momentum and torque and study the rotational versions of Newton's second law.

10–1—10–2 Generalizations of Angular Momentum and Torque

In Chapter 9 we treated the angular momentum of a rigid body in pure rotation. However, even a point particle moving with constant velocity has an angular momentum about a fixed point. This angular momentum is given by the most general expression for calculating angular momentum

$$\vec{L} = \vec{r} \times \vec{p},$$

where $\vec{r}$ is the position vector of the particle with respect to the point O about which the angular momentum is being calculated and $\vec{p}$ is its linear momentum. The notation used above is that of the **vector product** (or *cross product*) of the vectors $\vec{r}$ and $\vec{p}$.

The magnitude of the vector product is given by multiplying the magnitude of one vector by the perpendicular component of the other vector. That is,

$$L = rp_{\perp} = r_{\perp}p = rp\sin\theta,$$

where $p_{\perp}$ is the component of $\vec{p}$ perpendicular to $\vec{r}$, $r_{\perp}$ is the component of $\vec{r}$ perpendicular to $\vec{p}$, and θ is the smallest angle between the vectors $\vec{r}$ and $\vec{p}$. The direction of the resulting vector can be determined by noting two facts; **(a)** the vector $\vec{L}$ points along a line perpendicular to the plane formed by $\vec{r}$ and $\vec{p}$ (so it's perpendicular to each vector); and **(b)** the direction along this line is determined by the following right-hand rule for the vector product:

> *Point the fingers of your right hand along the direction of the first vector in the product. Orient your hand such that your fingers can curl toward the second vector through the smallest angle between them. Your extended thumb indicates which way along the perpendicular line the resultant vector points.*

It is important to notice that the order in which the vectors are multiplied matters; the magnitudes are the same but the directions are reversed,

$$\vec{A} \times \vec{B} = -\vec{B} \times \vec{A}.$$

There is also a way to determine the components of the vector that results from the vector product. The standard expression is

$$\vec{A} \times \vec{B} = \left(A_y B_z - A_z B_y\right)\hat{i} + \left(A_z B_x - A_x B_z\right)\hat{j} + \left(A_x B_y - A_y B_x\right)\hat{k}.$$

Just as the vector product can be used to express the most general form of the angular momentum of a particle about a point, it can also be used for the most general expression of the torque exerted by a force about a point O,

$$\vec{\tau} = \vec{r} \times \vec{F},$$

where $\vec{r}$ is the position vector from O to the point of application of the force. As when saw in Chapter 9, the magnitude of the torque is given by $\tau = rF\sin\theta$, and its direction is given by the same right-hand rule that we just stated; these facts are consistent with the definition of the vector product. It can easily be shown that the relationship between torque and angular momentum,

$$\vec{\tau}_{net} = \frac{d\vec{L}}{dt},$$

is still satisfied when these quantities are expressed in terms of the vector product.

Example 10–1 A person throws a 0.336-kg object vertically downward from a 7.83-m-high ledge with an initial speed of 2.25 m/s. Relative to a point O on the ground a distance $x = 5.92$ m from where it strikes, determine **(a)** the initial angular momentum of the object, **(b)** the torque due to its weight, and **(c)** the rate of change in angular momentum.

Setting it up: A diagram of the situation is helpful.
The information given in the problem follows:
Given: $m = 0.336$ kg , $h = 7.83$ m, $\vec{v}_i = -2.25\frac{m}{s}\hat{j}$,
$x = 5.92$ m; **Find**: **(a)** $\vec{L}_i$, **(b)** $\vec{\tau}$, **(c)** $d\vec{L}/dt$

Strategy:
There are two approaches that immediately come to mind for calculating the angular momentum and the torque. If we use $L = r_\perp p$ and $\tau = r_\perp F$ we can immediately see that $x = r_\perp$ for these cases. (If you don't see this, then extend the vertical line of $\vec{v}_i$ all the way down.) The magnitudes can then be calculated and the right-hand rules used to get the directions. The other choice is to use the cross product which gives both the magnitude and the direction all at once. For illustrative purposes, let us use the cross product.

Working it out:
(a) The initial angular momentum is given by $\vec{L}_i = \vec{r}_i \times \vec{p}_i = \vec{r}_i \times m\vec{v}_i = m(\vec{r}_i \times \vec{v}_i)$. We already know what the velocity vector is, and by inspection we can see that $\vec{r}_i = x\hat{i} + h\hat{j}$. Therefore,

$$\vec{L}_i = m\Big[(0-0)\hat{i} + (0-0)\hat{j} + (xv_y - 0)\hat{k}\Big] = xmv_y\hat{k}\,.$$

Inserting numerical values gives

$$\vec{L}_i = (5.92\,\text{m})(0.336\,\text{kg})(-2.25\tfrac{\text{m}}{\text{s}})\hat{k} = (-4.48\;\text{J}\cdot\text{s})\hat{k}\,.$$

(b) The torque due to its weight is given by $\vec{\tau} = \vec{r}_i \times \vec{W} = \vec{r}_i \times m\vec{g} = m(\vec{r}_i \times \vec{g})$. Thus, we have

$$\vec{\tau} = m\Big[(0-0)\hat{i} + (0-0)\hat{j} + (xg_y - 0)\hat{k}\Big] = xmg_y\hat{k}$$

Inserting numerical values gives

$$\vec{\tau} = (5.92\,\text{m})(0.336\,\text{kg})(-9.80\tfrac{\text{m}}{\text{s}^2})\hat{k} = (-19.5\;\text{N}\cdot\text{m})\hat{k}\,.$$

(c) To calculate the rate of change in angular momentum we seek an expression for the angular momentum as a function of time. From the result in part (a) we know that $\vec{L}_i = xmv_y\hat{k}$. For an object in free fall, the y-component of velocity is given by $v_y = v_i + g_y t$. Therefore, the angular momentum as it falls is

$$\vec{L}_i(t) = xm\big(v_i + g_y t\big)\hat{k}\,.$$

Taking the time derivative gives

$$\frac{d}{dt}\vec{L}_i(t) = xmg_y\hat{k}\,.$$

This is precisely the torque exerted by gravity (the weight of the mass). Therefore,

$$\frac{d}{dt}\vec{L}_i(t) = \vec{\tau} = (-19.5\;\text{N}\cdot\text{m})\hat{k}\,.$$

This result should make sense to you because the force of gravity is the only force on the mass and therefore exerts the only torque. So, this torque is the net torque.

What do you think? If you were to calculate the angular momentum and torque using $L_i = rp_{i\perp}$ and $\tau = rF_\perp$, what would be the values of $p_{i\perp}$ and $F_\perp$?

Practice Quiz

1. The position of a 0.68-kg particle is $\vec{r} = (1.6\;\text{m})\hat{i} - (4.1\;\text{m})\hat{j}$. It has a velocity of 5.7 m/s in the negative x direction. What is its angular momentum about the origin at that instant?
    (a) $(-3.9\;\text{J}\cdot\text{s})\hat{i}$ **(b)** $(-75\;\text{J}\cdot\text{s})\hat{k}$ **(c)** $(-29\;\text{J}\cdot\text{s})\hat{j}$ **(d)** $(-16\;\text{J}\cdot\text{s})\hat{k}$ **(e)** $(3.9\;\text{J}\cdot\text{s})\hat{i}$

2. If the y direction is vertically upward, what is the torque on the particle in question 1, about the origin, due to the gravitational force?
    (a) $(-11\;\text{N}\cdot\text{m})\hat{k}$ **(b)** $(-6.7\;\text{N}\cdot\text{m})\hat{j}$ **(c)** 0   **(d)** $(27\;\text{N}\cdot\text{m})\hat{i}$ **(e)** $(-1.1\;\text{N}\cdot\text{m})\hat{k}$

3. The axis of a rotating sphere is aligned along the north-south direction. If the sphere rotates clockwise as viewed from the southern side, what is the direction of its angular momentum?

(a) north   **(b)** south   **(c)** east   **(d)** west   **(e)** upward toward the sky

4. A uniform hoop of radius 0.75 m and mass 2.25 kg rotates with an angular velocity of 44 rad/s about an axis through its center. What is the magnitude of its angular momentum?

   (a) 0 J·s   **(b)** 74 J·s   **(c)** 1.3 J·s   **(d)** 33 J·s   **(e)** 56 J·s

10–3 The Dynamics of Rotation

In Chapter 8, we saw how the equations for the translational motion of a particle can be used to considerably simplify the behavior of entire systems of particles when the concept of the center of mass is used. Similarly, the expressions just studied for the angular momentum of a particle ($\vec{L} = \vec{r} \times \vec{p}$) and the torque on a particle due to a force ($\vec{\tau} = \vec{r} \times \vec{F}$) can often be used to simplify the rotational dynamics of entire systems when written in terms of the center of mass. However, there are times when the center of mass is not the best possible choice of reference point about which to calculate $\vec{L}$ and $\vec{\tau}$. Under these circumstances, there are several results that can be used for determining important dynamical quantities about an arbitrary point A if the quantity is known about the center of mass point.

Let $\vec{R}$ be a vector from the point A to the center of mass, $\vec{P} = M\vec{V}_{cm}$ be the total linear momentum of the system (with M being its total mass), and $\vec{F}_{tot}$ be the net external force on the system. Then, the angular momentum of the system with respect to A is

$$\vec{L}_A = \left(\vec{R} \times \vec{P}\right) + \vec{L}_{cm} .$$

The net torque on the system about A is

$$\vec{\tau}_A = \left(\vec{R} \times \vec{F}_{tot}\right) + \vec{\tau}_{cm} ,$$

where

$$\left(\vec{R} \times \vec{F}_{tot}\right) = \frac{d}{dt}\left(\vec{R} \times \vec{P}\right)$$

and

$$\vec{\tau}_{cm} = \frac{d}{dt}\vec{L}_{cm} .$$

Also useful for understanding the dynamics of rotational motion is the rotational analogue of the impulse of a force, namely, the **angular impulse** of a torque, $\vec{J}_\tau$. The average net torque acting on a system is given by $\vec{\tau}_{av} = \Delta\vec{L}/\Delta t$; so we have $\vec{\tau}_{av}\Delta t = \Delta\vec{L}$. Therefore, the angular impulse is

$$\vec{J}_\tau = \vec{\tau}_{av}\Delta t$$

and we have that

$$\vec{J}_\tau = \Delta\vec{L} .$$

Example 10–2 A solid cylinder, of mass 2.5 kg and radius 36 cm, is free to rotate about a vertical axis through its surface. A force of constant magnitude $F = 6.9$ N is applied tangentially to its surface at the center as shown. **(a)** Calculate the torque about the axis of rotation. **(b)** If it starts from rest, what is its angular momentum after 2.0 s?

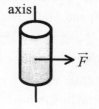

Setting it up: The diagram shows the cylinder with the axis along its back surface (not through the center). The applied force is tangent to the front surface causing it to have an angular acceleration. For the coordinate system, we take the axis to be along the z-direction with the x-direction coming out of the page and the y-direction to the right. The information given in the problem is the following:

Given: $m = 2.5$ kg, $R = 36$ cm, $F = 6.9$ N, $\omega_0 = 0$, $t = 2.0$ s

Find: **(a)** $\vec{\tau}$, **(b)** $\vec{L}$

Strategy:
Because the force is applied tangentially to the cylinder, we can calculate the magnitude of the torque by noting that the lever arm equals the diameter of the cylinder. The angular momentum after 2.0 s have elapsed can be determined using the angular impulse-momentum theorem.

Working it out:
(a) The magnitude of the torque is given by

$$\tau = r_\perp F = 2RF = 2(0.36 \text{ m})(6.9 \text{ N}) = 5.0 \text{ N} \cdot \text{m} \,;$$

The right-hand rule shows that the direction is along the positive z-axis.

(b) Over a period of time t, the angular impulse due to this constant torque is given by $\vec{J}_\tau = \vec{\tau} t$, which must equal the change in angular momentum $\Delta \vec{L}$. Because the cylinder starts from rest the initial angular momentum is zero. Therefore,

$$\vec{L} = \vec{\tau} t = (2RFt)\hat{k} = 2(0.36 \text{ m})(6.9 \text{ N})(2.0 \text{ s})\hat{k} = (9.9 \text{ J} \cdot \text{s})\hat{k} \,.$$

What do you think? Does the torque calculated above, about the axis through its surface, give the same result as the expression $\vec{\tau}_A = \left(\vec{R} \times \vec{F}_{\text{tot}}\right) + \vec{\tau}_{\text{cm}}$ does?

Practice Quiz

5. A solid circular disk can be connected such that it rotates about two different axes, one through a point C at its center and the other through a point P halfway between its center and the edge. If a force is applied tangentially to its rim as shown, about which axis will it have the largest angular acceleration?

    (a) the axis through C
    (b) the axis through P
    (c) it will be the same for both axes
    (d) There is not enough information to determine the answer.

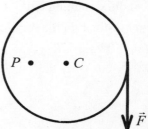

10–4 Conservation of Angular Momentum

Newton's second law for rotation says that the net torque on a system causes a change in the angular momentum of a system. Consequently, if there is no net torque on a system, there will be no change in the angular momentum; this is the principle of the conservation of angular momentum which was discussed briefly in Chapter 9.

An important example of a force that exerts no torque is a **central force**. Central forces are forces that are directed along the line from the source of the force to the object on which the force is applied. The most common and best known example is the gravitational force. Because gravity exerts no torque on objects moving under its influence, the angular momentum of these objects remains constant. The fact that the magnitude of the angular momentum is constant leads to **Kepler's second law** which states that the radius vector from the sun to a planet sweeps out equal areas, A, in equal times

$$\frac{dA}{dt} = \frac{L}{2m},$$

where L is the magnitude of the angular momentum and m is the mass of the planet. The fact that the direction of the angular momentum is constant demands that the orbit be restricted to a plane. It is worth noting that these same facts apply not just to planets orbiting the sun, but also to natural and artificial satellites orbiting planets, and to any object moving only under the influence of gravity (or any central force).

We often see the conservation of angular momentum at work in nonrigid bodies (such as people) as well. For such objects, some parts of the system can move relative to other parts, changing the rotational inertia of the system. Because $L = I\omega$, this change in I must be accompanied by a change in ω such that

$$I_0\omega_0 = I_f\omega_f.$$

This effect is often observed in sporting events such as gymnastics and diving in which athletes need to complete a certain number of rotations and speed up their rotation rate by tucking in their arms and legs.

Example 10–3 A merry-go-round with a moment of inertia of 750 kg·m^2 and a radius R of 2.55 m is rotating with an angular velocity of 9.42 rad/s clockwise (as viewed from above). A 334-N child runs at 2.76 m/s tangent to the rim of the merry-go-round and jumps onto it in the direction opposite its sense of rotation. With what angular velocity does the merry-go-round rotate after the child jumps onto it?

Setting it up:

The diagram is shown at the right. It shows a top view of the rotating merry-go-round and the running child's path before the child jumps on. The following information is given in the problem:

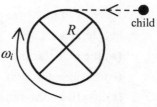

Given: $I = 750$ kg·m^2, $R = 2.55$ m, $\omega_i = 9.42$ rad/s, $Wt_c = 334$ N, $v_c = 2.76$ m/s

Find: ω_f

Strategy:

We can solve this problem by using the conservation of angular momentum. We will equate the total angular momentum before the child jumps on with the total angular momentum after the child jumps on.

Working it out:

Taking out of the page as the positive direction and into the page as the negative direction, the initial total angular momentum at the instant the child is about to jump on is

$$L_{total} = L_c - L_{MGR} = Rp_c - I_{MGR}\omega_i .$$

The final total angular momentum after the child jumps on is

$$L_{total} = I_{total}\omega_f = \left(I_{MGR} + m_c R^2\right)\omega_f .$$

The mass of the child is $m_c = Wt_c / g = 334 \text{ N}/(9.80 \text{ m/s}^2) = 34.08 \text{ kg}$.

We can now set the initial and final angular momenta equal to each other and solve for ω_f. This gives

$$Rp_c - I_{MGR}\omega_i = \left(I_{MGR} + m_c R^2\right)\omega_f$$

$$\therefore \quad \omega_f = \frac{Rm_c v_c - I_{MGR}\omega_i}{I_{MGR} + m_c R^2} .$$

Inserting numerical values gives

$$\omega_f = \frac{(2.55 \text{ m})(34.08 \text{ kg})(2.76\tfrac{m}{s}) - (750 \text{ kg} \cdot \text{m}^2)(9.42\tfrac{rad}{s})}{750 \text{ kg} \cdot \text{m}^2 + 34.08 \text{ kg}(2.55 \text{ m})^2} = -7.02 \text{ rad/s} .$$

What do you think? Would the final angular velocity be more or less if the child came from the opposite direction?

Practice Quiz

6. If the angular momentum of an object is conserved
    (a) it has zero net torque applied on it
    (b) it has constant angular velocity
    (c) it has constant angular acceleration
    (d) it has constant moment of inertia
    (e) The angular momentum of every object is conserved

7. If the angular momentum of a system of several objects is conserved,
    (a) there is a large net external torque on the system
    (b) no torque acts on any object in the system
    (c) the net external torque on the system is zero
    (d) the system rotates as a rigid body
    (e) none of the above

10–5 Work and Energy in Angular Motion

So far we have written rotational forms for several important quantities that we studied in translational motion, namely, displacement, velocity, acceleration, inertia, kinetic energy, force, and momentum. Two quantities that we have not addressed are work and power. These relationships are obtained most directly for the case of a rigid body rotating about a fixed axis. Recall that power is the rate at which energy is expended (or work is done) so that we can get the expression for the power requirements of a rotating object by differentiating the rotational kinetic energy $K = \frac{1}{2}I\omega^2 = \frac{1}{2}I\vec{\omega}\cdot\vec{\omega}$. Carrying out this differentiation gives for $P = dK/dt$

$$P = \vec{\tau}\cdot\vec{\omega}.$$

When a force gives rise to a torque that rotates an object, the force acts through a displacement and therefore does work. An analysis of this work, when the torque is along the axis, shows that, for infinitesimal displacements, it can be written in the rotational form $dW = \tau d\theta$. So that the overall work done is

$$W = \int \tau d\theta.$$

Rewriting this integral in terms of angular velocity also shows that for the rotational forms, the work-energy theorem still applies

$$W = \Delta K.$$

In Chapter 9, we saw that the kinetic energy of a rolling object can be written as the sum of the translational kinetic energy of the system treated as a single particle moving at the speed of the center of mass and the rotational kinetic energy of the system about the center of mass. In this section we find that this is, in fact, generally true for rigid body systems. Therefore, for a rigid body undergoing both translational and rotational motion, we have

$$K = \frac{1}{2}MV_{cm}^2 + \frac{1}{2}I_{cm}\omega^2.$$

Example 10–4 To sharpen a tool, the toolmaker uses a grinding wheel of radius 0.695 m with a coarse rim. The tool to be sharpened is pressed against the rim of the wheel with a force of 25.9 N (directed toward the center) as the wheel undergoes exactly 12 rotations. If the coefficient of kinetic friction between the tool and the rim is 0.644, how much work is done by friction in sharpening the tool?

Setting it up:

In this situation, torque is applied by the force of kinetic friction. This force acts tangentially along the rim of the grinding wheel. The following information is given in the problem:

Given: $F_N = 25.9$ N, $r = 0.695$ m, # of rev = 12, $\mu_k = 0.644$;   **Find:** W

Strategy:

The work done by friction can be treated in terms of the torque that the force of friction applies on the wheel about its center.

Working it out:

The torque about the center of the wheel is then given by

$$\tau = rf_k = r\mu_k N$$

The number of rotations of the wheel corresponds directly to the angular displacement:

$$\Delta\theta = 12 \text{ rev}\left(\frac{2\pi \text{ rad}}{\text{rev}}\right) = 75.40 \text{ rad}$$

Therefore, the work done by friction is

$$W = \tau\Delta\theta = r\mu_k N\Delta\theta = (0.695 \text{ m})(0.644)(25.9 \text{ N})(75.40 \text{ rad}) = 874 \text{ J}$$

What do you think? How would you calculate the work done without using the concept of torque?

Practice Quiz

8. If a constant torque of 5.2 N·m rotates an object through 136°, how much work is done by the torque?
   (a) 5.2 J   **(b)** 26 J   **(c)** 710 J   **(d)** 3.9 J   **(e)** 12 J

10–6 Collecting Parallels Between Rotational and Linear Motion

Through both Chapter 9 and this present chapter we have seen many similarities between how rotational and linear motion are treated. Here, we pause to glimpse at all the parallels between the two. One thing you will notice is that in most of the cases, the equations involving the rotational quantities are exactly the *same* as those involving the linear quantities except the names of the variables are different. This fact means that the mathematical techniques for analyzing the motion will be predominantly the same as well. So, to a large extent, mathematically, the study of rotational motion is just a repeat of your study of linear motion. However, the physics is different; and, of course, the physics is what's important.

<ins>Linear Motion</ins>	<ins>Rotational Motion</ins>
Infinitesimal displacement: $d\vec{r}$	Infinitesimal angular displacement: $\overrightarrow{d\theta}$
Velocity: $\vec{v} = \dfrac{d\vec{r}}{dt}$	Angular velocity: $\vec{\omega} = \dfrac{\overrightarrow{d\theta}}{dt}$
Acceleration: $\vec{a} = \dfrac{d\vec{v}}{dt}$	Angular acceleration: $\vec{\alpha} = \dfrac{d\vec{\omega}}{dt}$
Linear momentum: $\vec{p} = m\vec{v}$	Angular momentum: $\vec{L} = I\vec{\omega} = \vec{r} \times \vec{p}$
Force: $\vec{F}_{net} = \dfrac{d\vec{p}}{dt}$	Torque: $\vec{\tau}_{net} = \dfrac{d\vec{L}}{dt}$
Force: $\vec{F}_{net} = m\vec{a}$	Torque: $\vec{\tau}_{net} = I\vec{\alpha}$
Mass: m	Rotational inertia: $I = \sum mr^2$
Kinetic energy: $K = \frac{1}{2}mv^2$	Kinetic energy: $K = \frac{1}{2}I\omega^2$

$$\text{Impulse: } \vec{J} = \vec{F}_{av}\Delta t \qquad\qquad \text{Angular impulse: } \vec{J}_\tau = \vec{\tau}_{av}\Delta t$$

$$\text{Work: } W = \int \vec{F} \cdot d\vec{r} \qquad\qquad \text{Work: } W = \int \vec{\tau} \cdot \overline{d\theta}$$

$$\text{Power: } P = \vec{F} \cdot \vec{v} \qquad\qquad \text{Power: } P = \vec{F} \cdot \vec{\omega}$$

*10–7 Quantization of Angular Momentum

Everything said about angular momentum thus far has been said under the assumption that Newtonian physics applies. However, on atomic and subatomic scales, physics as determined by Newton's laws is no longer an accurate model of nature. Quantum physics tells us that ultimately, the magnitude of the angular momentum of a system can only have certain discrete values, that is, it is *quantized*. The quantization of angular momentum is such that components of angular momentum, such as, the z-component can only take on integer multiples of a certain constant, $\hbar \approx 10^{-34}\,\text{J}\cdot\text{s}$ (pronounced "*h*-bar"). In fact, not only is angular momentum quantized, but energy is as well.

*10–8 Precession

When an object rotates about an axis of symmetry with angular momentum $\vec{L}_z$, and is oriented such that there is a gravitational torque on the object that is perpendicular to $\vec{L}_z$, then the resulting rate of change in angular momentum $d\vec{L}/dt$ is such that the direction of $\vec{L}_z$ continually changes sweeping out a circular path. This rotation of the symmetry axis is called **precession**. It is useful to compare this to uniform circular motion studied in Chapter 3.

When an object undergoes uniform circular motion, the velocity is tangent to the path and the acceleration, the rate of change in velocity $d\vec{v}/dt$, is always perpendicular to it. This means that the magnitude of the velocity remains constant while its direction continually changes causing the velocity vector to rotate around a circular path. Similarly, precession is the circular motion of the angular momentum vector due to an orthogonal rate of change.

Example 10–5 A 3.0-kg horizontal beam that has a diameter of 10.0 cm and is exactly one meter long, sits on a pivot located 45 cm from one end of the beam. The beam is rotating about its central axis (as shown below) with an angular velocity of 50 rad/s. **(a)** In which direction does it precess, and **(b)** what is the angular frequency of precession?

Setting it up:

The diagram shows the beam rotating about its central axis such that it rotates counter clockwise as viewed from the right end. The information given in the problem is as follows:

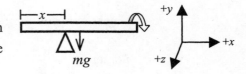

Given: $m = 3.0$ kg, $d = 10.0$ cm, $\omega_{central} = 50$ rad/s, $x = 45$ cm
Find: **(a)** direction, **(b)** ω_p

Strategy:

The direction of the precession is determined by the direction of the rate of change in angular momentum. This change in angular momentum is due to the torque exerted by the gravitational force. The component of angular momentum generated by the gravitational torque is accompanied by an angular velocity component, this new component of angular velocity is the angular frequency of the precession.

Working it out:

(a) Using the right-hand rule, the angular momentum about the central axis of the beam points in the $+x$ direction; $\vec{L}_{central} = I_{central}\omega_{central}\hat{i}$. While the torque from gravity about an axis through the pivot points in the $-z$ direction; $\vec{\tau}_{pivot} = -\ell mg\hat{k}$, where $\ell = 50$ cm $- x$ is the distance between the pivot axis and the center of mass. Therefore, at the instant shown, the direction of the change in angular momentum is in the $-z$ direction. This shows that the beam will rotate toward that direction, which means that it precesses counter clockwise as viewed from above.

(b) The magnitude of the net torque about the pivot is constant, $\tau_{pivot} = \ell mg = dL/dt$, representing the constant rate of change in the direction of the angular momentum vector. Thus, after a time dt, the angular momentum of the beam is

$$\vec{L} = \vec{L}_{central} + d\vec{L} = I_{central}\omega_{central}\hat{i} - \ell mg(dt)\hat{k}.$$

So, the central symmetry axis of the beam has rotated through an angular displacement

$$d\theta = \frac{dL}{L_{central}} = \frac{\ell mg(dt)}{I_{central}\omega_{central}}.$$

This fact means that

$$\omega_p = \frac{d\theta}{dt} = \frac{\ell mg}{I_{central}\omega_{central}}.$$

Now, the value of $I_{central}$ can be determined using Table 9.1 in your textbook. We require the rotational inertia of a solid cylinder about its central axis. We see from the table that its value is given by

$$I_{central} = \tfrac{1}{2}mr^2,$$

where the radius of the beam is half its given diameter, $r = 3.0$ cm. Therefore, including this in the equation for ω_p we now have

$$\omega_p = \frac{2\ell g}{r^2\omega_{central}} = \frac{2(0.05 \text{ m})(9.8 \text{ m/s}^2)}{(0.050 \text{ m})^2(50 \text{ rad/s})} = 7.8 \text{ rad/s}.$$

What do you think?   What would be different if the direction of $\omega_{central}$ were reversed?

Practice Quiz

9. When viewed from above, a spinning top rotates in the clockwise direction. It is then set on a horizontal floor such that its axis of rotation leans slightly. Which of the following statements is correct?

   (a) There will be no precession of the wheel.

   (b) The top will precess counter clockwise as viewed from above.

   (c) The top will precess clockwise as viewed from above.

   (d) The top will precess counter clockwise in the northern hemisphere and clockwise in the southern.

   (e) There is not enough information to know how the top will precess.

Reference Tools and Resources

I. Key Terms and Phrases

vector product a form of vector multiplication that produces a vector as the result

angular impulse a torque times the amount of time the force acts

central force a force that acts along the line from the source to the object experiencing the force

Kepler's second law the fact that the line from the sun to a planet sweeps out equal areas in equal times

precession the rotational motion of the axis of rotation for an object experiencing a torque perpendicular to the existing angular momentum

II. Important Equations

Name/Topic	Equation	Explanation
Angular momentum	$\vec{L} = \vec{r} \times \vec{p}$	The general definition of the angular momentum about a point.
Torque	$\vec{\tau} = \vec{r} \times \vec{F}$	The general definition of the torque about a point.
Angular impulse	$\vec{J}_\tau = \vec{\tau}_{av}\Delta t = \Delta \vec{L}$	The angular impulse momentum theorem.
Conservation of angular momentum	$I_0\omega_0 = I_f\omega_f$	The conservation of angular momentum for a nonrigid system of particles.
Power	$P = \vec{\tau} \cdot \vec{\omega}$	The power in terms of rotational quantities.
Work	$W = \int \tau d\theta$	The work done by a torque.

III. Know Your Units

Quantity	Dimension	SI Unit
Angular impulse ($\vec{J}_\tau$)	$\left[ML^2 T^{-1} \right]$	J·s

Practice Problems

1. What is the magnitude to one decimal place of the cross product of the vector: $21\hat{i} + 7\hat{j} + 40\hat{k}$ and the vector $10\hat{i} + 10\hat{j} + 10\hat{k}$?

2. What is the angle, to the nearest degree, between the two vectors in the preceeding problem?

3. A particle located (in meters) at $223\hat{i} + 57\hat{k}$ is subjected to a force (in newtons) of $73\hat{i}$. What is the magnitude of the torque to the nearest m·N?

4. By how much will the angular momentum, to the nearest kg·m^2/s, of the particle in the previous problem change in 9 seconds?

5. The angular momentum of a system of particles about their center-of-mass is given by $3t^2\hat{i} + 7t\hat{j} + 4\hat{k}$. What is its magnitude, to the nearest kg·m^2/s, if the time equals 10.1 s?

6. What is the magnitude of the net external torque, to the nearest tenth of a m·N, acting on the system of particles in the previous problem?

7. In this problem use 6×10^{24} kg and 6.4×10^6 m for the mass and radius of the Earth, respectively, and 7.3×10^{-5} radians/s for its angular velocity. What is the mass of an asteroid (as a fraction of the mass of the Earth to 5 decimal places) if it stops the Earth's rotation when it strikes the Earth tangentially at the equator traveling opposite to the Earth's rotation at a speed of 36 km/s?

8. A 437-g, 2-meter-long wooden rod is pivoted at its center. A 4-g bullet traveling at 252 m/s strikes it perpendicular to its length and half way between the center and the end. If the bullet exits with half its entrance velocity, what is the angular speed of the rod after the bullet leaves to the nearest thousandth of a radian/s?

9. A 1.36-kg wooden rod of length 0.66 m is hung from its end from a pivot on the ceiling. It is struck at its center of mass by a 28.1-g musket ball traveling at 40 m/s which embeds itself in the wood. What is the initial angular velocity of the rod to two decimal places?

10. In the previous problem, how high does the end of the rod rise to the nearest tenth of a centimeter?

Selected Solutions to End of Chapter Problems

7. For the projectile motion using the coordinate system shown, we find the position and velocity components:

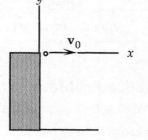

$$x = v_{ox}t = v_x t$$
$$y = v_{0y}t - \tfrac{1}{2}gt^2 = -\tfrac{1}{2}gt^2$$
$$v_y = v_{0y} - gt = -gt$$

When we express these as vectors, we have

$$\vec{r} = (v_x t)\hat{i} - \left(\tfrac{1}{2}gt^2\right)\hat{j}$$
$$\vec{p} = m\vec{v} = mv_x\hat{i} - mgt\,\hat{j}$$

The angular momentum is

$$\vec{L} = \vec{r}\times\vec{p} = \left(v_x t\hat{i} - \tfrac{1}{2}gt^2\hat{j}\right)\times\left(mv_x\hat{i} - mg\hat{j}\right) = -\tfrac{1}{2}gv_x t^2\hat{k}$$
$$= -\tfrac{1}{2}(0.060\text{ kg})(9.8\text{ m/s}^2)(25\text{ m/s})t^2\hat{k} = \boxed{-\left(7.4\text{ kg}\cdot\text{m}^2/\text{s}^3\right)t^2\hat{k}\text{ into the page}}$$

17. The torque, by definition, is the given by the following cross product:

$$\vec{\tau} = \vec{r}\times\vec{F} = \left(3\text{m}\,\hat{i} - \hat{j} - 5\text{m}\hat{k}\right)\times\left(2\text{m}\hat{i} + 4\text{m}\hat{j} + 3\text{m}\hat{k}\right) = \boxed{\left(17\hat{i} - 19\hat{j} + 14\hat{k}\right)\text{ N}\cdot\text{m}}$$

21. The location of the mass is given by $\theta = \omega t$. The torque is along the axle, into the page. Its magnitude is

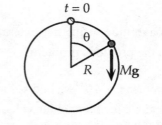

$$\tau = (R\sin\theta)Mg = \boxed{MgR\sin(\omega t)}.$$

33. Label the clay 1 and the wheel 2. The initial speed of the clay just before hitting the wheel is

$$v_1 = (2gh)^{1/2} = [2(9.8\text{ m/s}^2)(0.75\text{ m})]^{1/2} = 3.834\text{ m/s}.$$

Relative to the center of the wheel, the angular momentum of the clay-wheel system just before their collision is

$$L_i = m_1 v_1 R$$

with R the radius of the wheel. After the collision the rotational inertia of the system is

$$I_f = m_1 R^2 + \tfrac{1}{2}m_2 R^2$$

and so the angular speed of the system is obtained from conservation of angular momentum:

$$L_i = m_1 v_1 R = L_f = I_f\omega = (m_1 R^2 + \tfrac{1}{2}m_2 R^2)\omega$$

which gives

$$\omega = m_1 v_1/[(m_1 + \tfrac{1}{2}m_2)R] = (0.100\text{ kg})(3.834\text{ m/s})/[(0.100\text{ kg} + 10\text{ kg}/2)(0.50\text{ m})] = \boxed{0.15\text{ rad/s}}.$$

37. The work decreases the kinetic energy of the flywheel:

$$W = \Delta K = \tfrac{1}{2} I \left(\omega_f^2 - \omega_i^2 \right) \quad \Rightarrow \quad \omega_f = \sqrt{\omega_i^2 + \frac{2W}{I}}$$

Therefore,

$$\omega_f = \sqrt{\left(490 \text{ rad/s}\right)^2 + \frac{2\left(-1200 \text{ J}\right)}{0.033 \text{ kg} \cdot \text{m}^2}} = \boxed{490 \text{ rad/s}}$$

49. We choose the reference level for gravitational potential energy at the initial position. The kinetic energy will be the translational energy of the center of mass and the rotational energy about the center of mass. Because there is no work done by friction while the cylinder is rolling, for the work-energy theorem we have

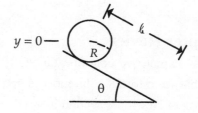

$$W_{\text{net}} = \Delta K + \Delta U$$
$$0 = \left(\tfrac{1}{2} M v^2 + \tfrac{1}{2} I \omega^2 - 0 \right) + Mg \left(0 - \ell \sin \theta \right)$$

Because the cylinder is rolling, $v = R\omega$. The rotational inertia is $\tfrac{1}{2} MR^2$. Thus we get

$$\tfrac{1}{2} M \left(R\omega \right)^2 + \tfrac{1}{2} \left(\tfrac{1}{2} MR^2 \right) \omega^2 = Mg\ell \sin \theta$$

which we write as

$$\omega = \sqrt{\frac{4g\ell \sin \theta}{3R^2}} = \sqrt{\frac{4\left(9.8 \text{ m/s}^2\right)\left(1.5 \text{ m}\right)\left(\sin 28^\circ\right)}{3(0.042 \text{ m})^2}} = \boxed{72 \text{ rad/s}}$$

Answers to Practice Quiz

1. (d) 2. (a) 3. (a) 4. (e) 5. (c) 6. (a) 7. (c) 8. (e) 9. (d)

Answers to Practice Problems

1. 405.7   2.   31°

3. 4,161 m·N   4.   37,449 kg·m²/s

5. 314 kg·m²/s   6.   61.0 m·N

7. 0.00519   8.   1.730 rad/s

9. 1.85 rad/s   10. 2.5 cm

CHAPTER 11

STATICS

Chapter Objectives

After studying this chapter, you should

1. be able to apply the conditions for static equilibrium.
2. be able to analyze simple cases of longitudinal, volume, and shear stress.

Chapter Review

In this chapter, we continue the study of rotation by looking at the dynamics of rotational motion. It is here that we introduce the most general forms of angular momentum and torque and study the rotational versions of Newton's second law.

11–1—11–3 Static Conditions for Rigid Bodies and Applications

In many applications, especially in the design of buildings and other approximately rigid structures, objects need to be balanced so that they do not fall down or tip over. This means that the objects must be in equilibrium. To be in complete equilibrium an object must be in both translational equilibrium and rotational equilibrium. A rigid body in translational equilibrium has zero net external force acting on it and therefore its center of mass moves with constant velocity. A rigid body in rotational equilibrium has zero net external torque acting on it and therefore it rotates about its center of mass with constant angular velocity.

An important special case of equilibrium occurs when the constant translational and rotational velocities are zero; $\vec{V}_{cm} = 0$ and $\vec{\omega}_{cm} = 0$. This special case is referred to as **static equilibrium**. The primary conditions for static equilibrium, however, are those concerning the net external force and torque on the body:

$$\sum \vec{F}_i = 0, \quad \sum \vec{\tau}_i = 0$$

When doing an analysis for equilibrium it is useful to remember that the center-of-mass concept is often very useful here. The weight of an object can exert a torque about any axis of rotation that does not pass through the center of mass. Therefore, with regard to torque, we can treat rigid objects like point masses whose mass is located at the center of mass. Let us summarize a useful approach for performing an analysis of static equilibrium.

1. Draw a free body diagram of the object(s) in equilibrium. You may need to *mentally* apply the equilibrium conditions to determine the directions for all the forces. In particular, the condition for the net torque may be applied about several axes to get all the directions correct. If you are unable to determine all the directions with certainty, guess the one(s) you don't know and proceed; your solution should provide you with the correct direction as well as the magnitude.

2. With a completed free body diagram, choose *one* axis for calculating *all* the torques in the equation that you will use to find your numerical solution. If the problem has more than one unknown force acting at different locations, it is recommended that you choose the axis at the location of one of the unknown forces. Such a choice simplifies the required mathematical solution.

3. Write out the equations for the equilibrium conditions for the forces and the torques.

4. Solve these equations for the desired information.

Example 10–1 Consider the attempt to hang a decorative ornament in Example 9-3. However, this time the cord is strong enough to support the structure. The horizontal beam is 82.3 cm long and weighs 12.1 N. The ornament weighs 7.62 N. **(a)** What is the tension in the cord? **(b)** What are the horizontal and vertical components of the force that the wall directly applies to the beam?

Setting it up:

The diagram is shown at the right. The structure maintains static equilibrium. The following information is given in the problem:

Given: $L = 82.3$ cm, $W_b = 12.1$ N, $W_o = 7.62$ N

Find: (a) T, (b) $F_{wall,x}$ and $F_{wall,y}$

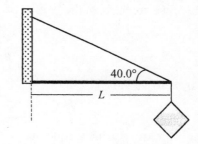

Strategy:

Our strategy will follow the four steps outlined above.

Working it out:

We start with the free body diagram of the beam shown below.

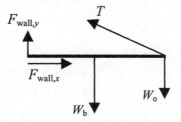

Let's discuss this diagram. Notice that the force exerted by the wall is divided into x and y components, whereas the tension is not. This was done because the problem asks for the full magnitude of $\vec{T}$ and the individual components of $\vec{F}_{wall}$. The point is, in your free body diagrams, you are free to represent any force either by its components or by a single vector in its direction – just don't do both for the same force.

The direction of $\vec{T}$ is given in the diagram of the problem and the weights are always downward. To determine the directions of the x and y components of $\vec{F}_{wall}$ we need to mentally apply the equilibrium conditions. We know that $F_{wall,x}$ points to the right because the only other force in the x direction is from the tension in the cord which clearly pulls to the left. If we try to determine $F_{\text{wall},y}$ using equilibrium of forces, we have a problem because while the weights act downward, the tension acts upward and we don't know what the tension is yet preventing us from saying for sure which way $F_{\text{wall},y}$ points. However, if we mentally apply the condition for torques about an axis at the right end of the beam, we see that $\vec{T}$ and $\vec{W}_o$ will not apply any torque because they have zero lever arm from an axis located there. This leaves nonzero torque from $\vec{W}_b$ and $F_{\text{wall},y}$ only which must cancel each other. With $\vec{W}_b$ being downward, it applies a torque directed out of the page about that axis. Therefore, $F_{\text{wall},y}$ has to be oriented such that it applies a torque into the page about an axis through the right end. Therefore, by the right-hand rule, we see that $F_{\text{wall},y}$ must be upward.

Now that we have the free-body diagram set, we can choose the axis to be located at the left end of the beam, where the force from the wall is exerted. The equilibrium conditions for the forces are

$$\sum F_x = F_{\text{wall},x} - T_x = 0$$
$$\sum F_y = F_{\text{wall},y} + T_y - W_b - W_o = 0.$$

For the torques, we have

$$\sum \tau_i = T\ell_T - W_b \tfrac{L}{2} - W_o L = 0,$$

where the lever arm for the torque due to the tension in the cord is shown below

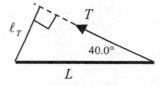

From the diagram, we can see that $\ell_T = L\sin(40.0°)$.

(a) The value of choosing the axis at the unknown force F_{wall} is now apparent because we only have one unknown in the equation for the equilibrium of torques. Therefore, we can now solve this equation for that unknown (the tension). Substituting in the expression for the lever arm, we have

$$TL\sin(40.0°) - W_b \tfrac{L}{2} - W_o L = 0 \quad \Rightarrow \quad T\sin(40.0°) - \tfrac{1}{2}W_b - W_o = 0$$

$$\therefore \quad T = \frac{W_o + W_b/2}{\sin(40.0°)} = \frac{7.62 \text{ N} + (12.1 \text{ N})/2}{\sin(40.0°)} = 21.3 \text{ N}.$$

(b) Now that the tension is known, we can use the equilibrium conditions for the forces to determine the components of the force exerted by the wall.

$$\sum F_x = F_{\text{wall},x} - T_x = 0 \quad \therefore \quad F_{\text{wall},x} = T_x = T\cos(40.0°)$$

$$\therefore \quad F_{\text{wall},x} = (21.27 \text{ N})\cos(40.0°) = 16.3 \text{ N}.$$

$$\sum F_y = F_{wall,y} + T_y - W_b - W_o = 0 \quad \therefore \quad F_{wall,y} = W_b + W_o - T_y = W_b + W_o - T\sin(40.0°)$$

$$\therefore \quad F_{wall,y} = 12.1 \text{ N} + 7.62 \text{ N} - (21.27 \text{ N})\sin(40.0°) = 6.05 \text{ N}$$

What do you think? Why doesn't the length of the beam matter in this calculation?

Practice Quiz

1. If both the net force and the net torque on an object is zero, the object will
    (a) remain motionless
    (b) have constant angular acceleration
    (c) have constant linear acceleration
    (d) have constant angular velocity
    (e) have no forces applied to it

2. A 1.5-m-long, uniform beam weighing 5.5 N is supported at one end by a fulcrum as shown and at the opposite end by a vertical force $\vec{F}$. What is the lever arm for the torque applied by $\vec{F}$ about an axis through the point of contact between the fulcrum and the beam.
    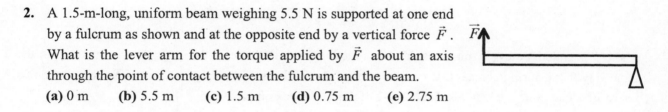
    (a) 0 m   **(b)** 5.5 m   **(c)** 1.5 m   **(d)** 0.75 m   **(e)** 2.75 m

3. For the situation described in question 2, what is the lever arm for the torque due to the weight of the beam about an axis through the point of contact between the fulcrum and the beam?
    (a) 0 m   **(b)** 5.5 m   **(c)** 1.5 m   **(d)** 0.75 m   **(e)** 2.75 m

4. For the situation described in question 2, what is the lever arm for the torque due to the force applied by the fulcrum about an axis through the point of contact between the fulcrum and the beam?
    (a) 0 m   **(b)** 5.5 m   **(c)** 1.5 m   **(d)** 0.75 m   **(e)** 2.75 m

5. For the situation described in question 2, what is the magnitude of the force $\vec{F}$ so that the beam remains in the static equilibrium shown?
    (a) 2.8 N   **(b)** 5.5 N   **(c)** 3.0 N   **(d)** 0.75 N   **(e)** 8.2 N

6. For the situation described in question 2, what is the magnitude of the force applied by the fulcrum?
    (a) 2.8 N   **(b)** 5.5 N   **(c)** 3.0 N   **(d)** 0.75 N   **(e)** 8.2 N

11–4 Solids and How They Respond to Forces

The intermolecular forces at work within a solid give rise to the "springlike" behavior of solids when they are deformed by external forces. When a force of magnitude F is applied perpendicularly to a side of an object of area A so as to stretch the object, we say that the object is under **tension**; if this force is applied in such a way as to squash the object, we say that it is under **compression**. The force responsible for this tension or compression is said to apply a longitudinal **stress** to the object, S, which is given by the force per unit area

$$S = \frac{F}{A}.$$

The usual convention is that a tension causes a positive stress and a compression causes a negative stress.

When an object is under a longitudinal stress it responds to this stress by a change in length along the direction of the force. A measure of this deformation is given by the fractional change in length, called the compressional **strain** of the object, e

$$e = \frac{\Delta L}{L},$$

where L is the original length of the object along the direction of the force. When the compressional strain is small, the intermolecular forces behave more like ideal springs than when the strain is large. Therefore, under the conditions of small strains, there is a direct proportionality between the stress and the strain

$$\frac{S}{e} = Y,$$

where Y is called **Young's modulus**. Young's modulus is a measure of the stiffness of a material in the same way that a spring constant, k, is a measure of the stiffness of a spring. Values of Young's modulus for various materials are listed in Table 11–1 of your textbook on page 326.

Besides a compressional strain, a longitudinal stress also deforms an object along the dimensions transverse to the direction of the force. That is, stretching an object also makes it thinner and squashing an object makes it fatter. The fractional changes in width h and depth w are directly proportional to the compressional strain, by a positive factor, σ, called *Poisson's ratio*

$$\frac{\Delta h}{h} = \frac{\Delta w}{w} = -\sigma e.$$

The value of Poisson's ratio depends on the material.

Deformations that change the entire volume of an object, such as when fluid pressure is applied perpendicularly over the entire surface of a completely submerged object, obey a relation similar to that of the longitudinal stress and the resulting strain. For this case, the *volume stress* is $p = F / A$, where A is the total suface area of the object. The resulting *volume strain* is $e_V = \Delta V / V$. For small strains, the stress and strain are directly proportional to each other; the proportionality factor is called the **bulk modulus** B,

$$-\frac{p}{e_V} = B.$$

The minus sign expresses the convention that inward pressures are positive; because the result of an inward pressure is a decrease in volume ($\Delta V < 0$), the minus sign ensures that the ratio is always positive.

Another form of deformation (a *shear* deformation) occurs when the applied force, $F_{\parallel}$, is directed along the surface of the area over which it is applied; let's call it the top surface for convenience. In this case, if the opposite side of the object, the bottom surface, is held fixed, the shape of the object will change because the top surface will extend a distance ΔL beyond the bottom surface. If the distance between the top and bottom surfaces is L, then we have a shear stress of $\Delta L/L$, as the result of the shear strain $F_\parallel/A$. The proportionality factor is called the **shear modulus**, G

$$\frac{F_\parallel / A}{\Delta L / L} = G .$$

Example 11–2 An aluminum cube has an edge length of 5.00 cm. If the bulk modulus of aluminum is 7.00×10^{10} N/m^2 , what pressure is required to decrease its size such that it has an edge length of 4.75 cm?

Setting it up: The information given in the problem is the following:
Given: $L_0 = 5.00$ cm, $L = 4.75$ cm, $B = 7.00 \times 10^{10}$ N/m^2
Find: p

Strategy:
To solve this problem, we need to determine the volume strain in the aluminum block. Then, the relationship between volume stress and strain can be used.

Working it out:
The volume strain is given by

$$e_V = \frac{\Delta V}{V} = \frac{V_f - V_0}{V_0} = \frac{(4.75 \,\text{cm})^3 - (5.00 \,\text{cm})^3}{(5.00 \,\text{cm})^3} = -0.1426 .$$

The required pressure is given by

$$p = -e_V B = -(-0.1426)\left(7.00 \times 10^{10} \,\text{N/m}^2\right) = 9.98 \times 10^9 \,\text{Pa}$$

What do you think? Is this a large pressure or not?

Practice Quiz

7. A certain pressure p decreases the volume of an object by an amount ΔV. If the object's original volume was twice the magnitude of ΔV, its volume strain would be
   (a) –1/2   (b) –2   (c) –4   (d) –1/4   (e) –1

8. A steel beam of radius 8.5 cm is normally 1.5 m long. If it is used to support a weight of 500 N on its end, what will be its change in length?
   (a) 2.1 mm   (b) 1.5 mm   (c) 1.6 m   (d) 8.5 cm   (e) 0.16 μm

Reference Tools and Resources

I. Key Terms and Phrases

static equilibrium the state of motion in which an object neither translates nor rotates

tension a force that acts to stretch an object

compression a force that acts to compress an object

stress the applied force per unit area that deforms a substance

strain the deformation that results from a stress applied to an object

Young's modulus the proportionality factor between small longitudinal stress and strains

bulk modulus the proportionality factor between small volume stresses and strains

shear modulus the proportionality factor between small shear stresses and strains

II. Important Equations

Name/Topic	Equation	Explanation
Equilibrium conditions	$\sum \vec{F_i} = 0, \ \sum \vec{\tau_i} = 0$	The general conditions for equilibrium.
Stress and strain	$\dfrac{S}{e} = Y$	The relationship between small compressional strain and longitudinal stress.
Stress and strain	$\dfrac{\Delta h}{h} = \dfrac{\Delta w}{w} = -\sigma e$	The relationship between small compressional strain and the deformation in the transverse dimensions.
Stress and strain	$-\dfrac{p}{e_v} = B$	The relationship between small volume strains and volume stresses.
Stress and strain	$\dfrac{F_{\parallel}/A}{\Delta L/L} = G$	The relationship between small shear strains and shear stresses.

III. Know Your Units

Quantity	Dimension	SI Unit
Stress (F/A, p)	$[ML^{-1}T^{-2}]$	N/m^2
Strain (e)	dimensionless	—
Modulus (Y, G, B)	$[ML^{-1}T^{-2}]$	N/m^2

Practice Problems

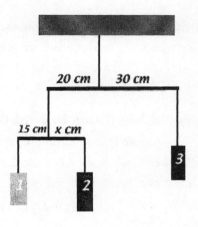

1. In the mobile at the right $m_1 = 0.84$ kg and $m_2 = 0.76$ kg. If the masses are to be balanced, what must the unknown distance be, to the nearest tenth of a cm?

2. To the nearest hundredth of a kilogram, in the mobile above, what is the value for m_3?

3. A 76-kg student sits on a chair which is solely supported by a solid 0.64-meter long steel rod 1 cm in diameter. To the nearest micron (millionth of a meter), what is the change in length of the rod produced by the student's weight?

4. The steel rod in the previous problem has a tensile strength of 564 MN/m². To the nearest hundredth of a millimeter, what is the minimum radius of the supporting rod that will support the student?

5. In the figure to the right, the bar is 2.7 meters long and has a mass of 115 kg. To the nearest newton, what is the magnitude of the hinge force?

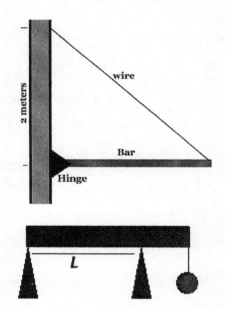

6. A large uniform "butcher block," the lower figure at the right, rests on two supports and has a weight hanging from its end. The block has a mass of 100 kg and a length of 2 meters. The distance L is 1.35 meters and the hanging weight is 151 newtons. To the nearest newton, what is the force on the left support?

7. In the previous problem, what is the force on the right support?

8. In Problem 6 the hanging weight is 167 newtons. To the nearest hundredth of a meter, what is the minimum value for L for the configuration to remain stable?

9. In Problem 6, L is 1.45 meters. To the nearest tenth of a kilogram, if the configuration is to remain stable, what is the minimum value for the hanging weight?

10. A ladder can fall for two reasons, if it is set too steep and the climber gets his mass to the left of the ladder's base, it is likely that the ladder will fall over backwards. If the ladder is set at too shallow of an angle the required force of friction between the ladder and the ground might be too great and the base of the ladder will slip. Assume that there is no friction between the ladder and the wall and that

the ladder is effectively weightless. The coefficient of friction between the base of the ladder and the ground is 0.23. The person using the ladder will be 3/4 of the way up the ladder. If the person climbing the ladder has a weight of 980 newtons and the ladder is 3.02 meters long, how far from the wall can the base of the ladder be placed and not slip, to the nearest hundredth of a meter?

Selected Solutions to End of Chapter Problems

3. We choose the coordinate system shown, with positive torques clockwise. We write $\Sigma\tau = I\alpha$ about the point A from the force diagram for the board:

$$\sum \tau_A = MgD - F_{N2}L = 0$$

which gives

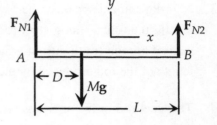

$$F_{N2} = Mg(D/L) = (24\,\text{kg})(9.8\tfrac{\text{m}}{\text{s}^2})(0.9\,\text{m})/(2.2\,\text{m}) = 96\,\text{N}$$

We write $\Sigma F_y = ma_y$ from the force diagram for the board and worker:

$$F_{N1} + F_{N2} - Mg = 0$$

so that

$$F_{N1} = Mg - F_{N2} = (24\,\text{kg})(9.8\tfrac{\text{m}}{\text{s}^2}) - 96\,\text{N} = 139\,\text{N}$$

The forces on the workmen are the reactions to these normal forces, therefore,

$$\boxed{139\,\text{N down}} \quad \text{and} \quad \boxed{96\,\text{N down}}.$$

19. We choose the coordinate system shown, with positive torques counterclockwise.

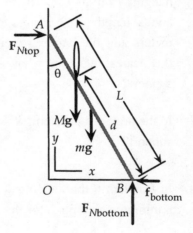

(a) We write $\Sigma\tau = I\alpha$ about the point B from the force diagram for the ladder and man:

$$\Sigma\tau_B = mg(\tfrac{1}{2}L)\sin\theta + Mgd\sin\theta - F_{N\,top}L\cos\theta = 0.$$

We write $\Sigma F_x = ma_x$ from the force diagram for the ladder and man:

$$F_{N\,top} - f_{bottom} = 0.$$

We write $\Sigma F_y = ma_y$ from the force diagram for the ladder and man:

$$F_{N\,bottom} - (m+M)g = 0.$$

As the man climbs the ladder, the friction force at the bottom increases. At the point where slipping begins,

$$f_{bottom} = f_{bottom,max} = \mu_s F_{N\,bottom} = \mu_s(m+M)g.$$

Then we have

$$F_{N\,top} = \mu_s(m+M)g.$$

Using these in the torque equation, we get

$$d = \frac{\mu_s (m + M) L \cos\theta - m(L/2)\sin\theta}{M \sin\theta}$$

$$= \frac{0.40(10 \text{ kg} + 80 \text{ kg})(4 \text{ m})\cos 30° - 10 \text{ kg}\left(\frac{4 \text{ m}}{2}\right)\sin 30°}{(80 \text{ kg})\sin 30°}$$

$$= \boxed{2.9 \text{ m}}$$

(b) We write $\Sigma\tau = I\alpha$ about the point A from the force diagram for the ladder and man:

$$\Sigma\tau_A = F_{N\,\text{bottom}}(L \sin\theta) - Mg(L - d) \sin\theta - mg(\tfrac{1}{2} L) \sin\theta - f_{\text{bottom}}L \cos\theta = 0.$$

We write $\Sigma\tau = I\alpha$ about the point O from the force diagram for the ladder and man:

$$\Sigma\tau_O = F_{N\,\text{bottom}}(L \sin\theta) - Mg(L - d) \sin\theta - mg(\tfrac{1}{2} L) \sin\theta - F_{N\,\text{top}}L \cos\theta = 0.$$

We can combine these three torque equations to obtain the force equations. If we add the equations for points A and O, we get

$$F_{N\,\text{top}} = f_{\text{bottom}}$$

which is the x-equation. If we subtract the equations for points B and O, we get

$$F_{N\,\text{bottom}} = \mu_s(m + M)g$$

which is the y-equation.

35. We choose the coordinate system shown, with positive torques clockwise. We write $\Sigma\tau = I\alpha$ about the point B from the force diagram for the beam:

$$\Sigma\tau_B = mg(\tfrac{1}{2} L \cos\theta) + Mg[(L - d) \sin\theta]$$
$$- T\cos\phi (L \sin \theta) - T \sin\phi (L \cos \theta) = 0,$$

which gives

$$M = \{[TL(\cos\phi \sin\theta - \sin\phi \cos\theta)/g]$$
$$- \tfrac{1}{2} mL \cos\theta\} / [(L - d) \cos\theta].$$

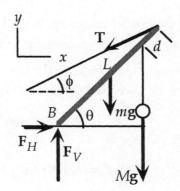

We see that maximum M corresponds to maximum T, so we have

$$M_{\text{max}} = \frac{\left[(10\times10^3\,\text{N})(3.00 \text{ m})\frac{(\cos 30° \sin 45° - \sin 30° \cos 45°)}{9.8 \text{ m/s}^2}\right] - \tfrac{1}{2}(100 \text{ kg})(3.00 \text{ m})\cos 45°}{\left[(3.00 \text{ m} - 0.25 \text{ m})\cos 45°\right]} = \boxed{353 \text{ kg}}$$

47. From the relation between stress and strain, we have

$$\text{stress} = Y (\text{strain}) \quad \Rightarrow \quad mg/\pi r^2 = Y(\Delta L/L)$$

which gives

$$\Delta L = \frac{mgL}{\pi r^2 Y} = \frac{(800\text{ kg})(9.8\text{ m/s}^2)(250\text{ m})}{\pi(1.0\times10^{-2}\text{ m})^2(21\times10^{10}\text{ N/m}^2)} = 2.97\times10^{-2}\text{ m} = \boxed{2.97\text{ cm}}$$

55. We call the length of the beam L_{beam}. The initial length of the cable is

$$L_0 = L_{\text{beam}}/\cos30° = 2L_{\text{beam}}/\sqrt{3}$$

and the distance along the wall from the beam to the cable is

$$D = L_{\text{beam}}\tan30° = L_{\text{beam}}/\sqrt{3}.$$

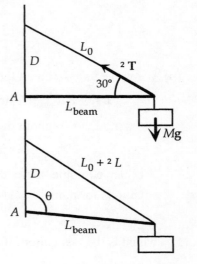

When the load is hung at the end of the beam, there must be an additional tension in the cable to maintain equilibrium:

$\Sigma\tau_A = (\Delta T \sin 30°)L_{\text{beam}} - MgL_{\text{beam}} = 0$, or

$\Delta T = Mg/\sin30° = (30\text{ kg})(9.8\text{ m/s}^2)/\sin 30° = 588$ N,

which causes an additional stress:

$$\frac{\Delta T}{\pi r^2} = 588\text{ N}/[\pi(1\times 10^{-3}\text{ m})^2] = 1.87\times 10^8\text{ N/m}^2.$$

(Even when the initial tension is added, this is less than the tensile strength, so the cable does not break.) The additional stress produces an elongation:

$$\Delta L/L_0 = \Delta T/AY.$$

This elongation causes the beam to drop below the horizontal (exaggerated in the diagram). We find the angle θ from a geometrical formula for a triangle:

$$D^2 + L_{\text{beam}}{}^2 - 2DL_{\text{beam}}\cos\theta = (L_0 + \Delta L)^2 = L_0{}^2(1 + \Delta L/L_0)^2 \approx L_0{}^2(1 + 2\Delta L/L_0),$$

where we have used the fact that $\Delta L \ll L_0$. Because $D^2 + L_{\text{beam}}{}^2 = L_0{}^2$, this becomes

$$-2(L_{\text{beam}}/\sqrt{3})L_{\text{beam}}\cos\theta = (2L_{\text{beam}}/\sqrt{3})^2\, 2\, \Delta L/L_0,$$

which reduces to

$\cos\theta = -(4/\sqrt{3})(\Delta L/L_0) = -(4/\sqrt{3})(\Delta T/AY) = -(4/\sqrt{3})(588\text{N})/[\pi(1\times10^{-3}\text{m})2(2.1\times10^5\frac{\text{MN}}{\text{m}^2})] = 0.00206,$

which gives $\theta = 90.12°$. Thus the beam is $\boxed{0.12°\text{ below the horizontal}}$.

65. We choose the coordinate system shown, with positive torques counterclockwise.

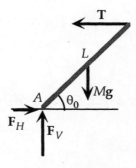

(a) We write $\Sigma\tau = I\alpha$ about the point A from the force diagram for the beam:

$$\Sigma\tau_A = Mg(\tfrac{1}{2}L)\cos\theta_0 - TL\sin\theta_0 = 0,$$

which gives

$$T = \tfrac{1}{2}\,\boxed{Mg\cot\theta_0}.$$

(b) When the cable snaps, the tension is zero, so we have

$$\Sigma\tau_A = I\alpha$$
$$Mg(\tfrac{1}{2}L)\cos\theta_0 = \tfrac{1}{3}ML^2\alpha,$$

which gives $\alpha = \boxed{(3g\cos\theta_0)/2L}$.

(c) The angular acceleration is not constant. Rather than integrate, we use the work-energy theorem, with the reference level for potential energy at the horizontal position. No work is done by the force at the pivot, so we have

$$W = \Delta K + \Delta U$$
$$0 = (\tfrac{1}{2}I_A\omega^2 - 0) + Mg(0 - \tfrac{1}{2}L\sin\theta_0).$$

When we use $I_A = \tfrac{1}{3}ML^2$, we get

$$\boxed{\omega = \sqrt{(3g\sin\theta_0)/L}}$$

Answers to Practice Quiz

1. (d) **2.** (c) **3.** (d) **4.** (a) **5.** (a) **6.** (a) **7.** (a) **8.** (e)

Answers to Practice Problems

1.	16.6 cm	**2.**	1.07 kg
3.	29 μm	**4.**	0.65 mm
5.	947 N	**6.**	182 N
7.	950 N	**8.**	1.15 m
9.	802.6 kg	**10.**	0.89 m

CHAPTER 12

GRAVITATION

Chapter Objectives

After studying this chapter, you should

1. be able to apply Newton's universal law of gravitation.
2. be able to understand and use Kepler's laws of planetary motion.
3. be able to determine the gravitational potential energy of a collection of bodies.
4. be able to determine the type of orbit an object will have based on its total energy.
5. be able to calculate the acceleration due to gravity at and above the surface of a planet.

Chapter Review

In this chapter, we study **gravity** — a fundamental force of nature. Of the four fundamental forces, gravity is the most familiar. The other fundamental forces will be discussed later in the text.

12–1—12–3 Newton's Universal Law of Gravitation & The Motion of Planets and Satellites

Gravitation is the phenomenon that between every two objects there is a force of attraction. **Newton's law of universal gravitation** describes the behavior of this force. Between any two point masses m_1 and m_2, the magnitude of the gravitational force on each mass due to the other is given by

$$F = G \frac{m_1 m_2}{r^2}$$

where r is the distance between the two masses, and G is a constant called the *universal gravitational constant*. The value of this constant is

$$G = 6.67 \times 10^{-11} \ \text{N} \cdot \text{m}^2/\text{kg}^2 .$$

The force on each mass points directly at the other mass because each mass attracts the other toward it. The small magnitude of the constant G explains why we only notice the effects of gravity when at least one of the masses is very large, such as with the mass of a planet or the sun.

For cases in which one mass is much larger than the other, as with planets and comets orbiting the Sun and artificial satellites orbiting Earth, the larger mass, $M \gg m$, is effectively stationary. Therefore, its center can be taken as the origin of a (polar) coordinate system in which the motion of the smaller mass is described. If r is taken as the radial coordinate of m relative to M, then Newton's law of gravitation can be written in vector form as

$$\vec{F} = -\frac{GMm}{r^2} \hat{r} ,$$

where the minus sign ensures that the force on m is toward M.

Newton's law of universal gravitation, together with his three laws of motion, can be used to explain **Kepler's three laws of planetary motion**.

1. Planets move in planar elliptical orbits, with the Sun at one focus of the ellipse.

2. The radius vector from the sun to the planet sweeps out equal areas in equal times.

3. If T is the time it takes for a planet to make one full revolution around the Sun, and R is half the major axis of the ellipse, then the ratio T^2/R^3 is the same for all planets:

$$\frac{T^2}{R^3} = \frac{4\pi^2}{GM}.$$

A couple of things to keep in mind about Kepler's laws is that while the Sun is at one focus of a planet's orbit, the other focus is empty; and that circular orbits are also possible because a circle is a form of an ellipse in which both the major and minor axes are equal. While Kepler's laws focus on the planetary orbits, it is worth noting that, under the influence of gravity, other paths are possible. The possible paths fall under the category of the *conic sections*. These paths are the *circle*, the *ellipse*, the *parabola*, and the *hyperbola*.

Example 12–1 If it is required to have a satellite in a circular orbit with an orbital period of exactly two days, at what altitude h above the ground should it be placed in orbit?

Setting it up:

The diagram is shown at the right. The sketch shows Earth and an artificial satellite a height h above Earth's surface. The given information is

Given: $T = 2$ d; **Find:** h

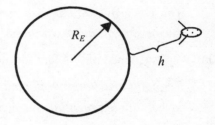

Strategy:

For circular orbits, the semimajor axis is the same as the radius. This radius extends from the center of the Earth to the satellite. We can use Kepler's third law to relate the period to the radius of the orbit.

Working it out:

Writing the radius of the orbit as $R = R_E + h$, the equation for Kepler's third law gives

$$\frac{T^2}{R^3} = \frac{4\pi^2}{GM} \quad \Rightarrow \quad \left(R_E + h\right)^3 = \frac{GMT^2}{4\pi^2}.$$

Solving this for h, we get

$$h = \left(\frac{GMT^2}{4\pi^2}\right)^{1/3} - R_E.$$

You can find values for the mass and radius of the Earth in the inside front cover of your textbook. The given orbital period should be converted to seconds.

$$T = 2\text{ d}\left(\frac{24\text{ h}}{\text{d}}\right)\left(\frac{3600\text{ s}}{\text{h}}\right) = 1.728 \times 10^5\text{ s}.$$

Now, we can find the altitude to be

$$h = \left[\frac{\left(6.67 \times 10^{-11} \, \frac{N \cdot m^2}{kg^2}\right)\left(5.976 \times 10^{24} kg\right)\left(1.728 \times 10^5 s\right)^2}{4\pi^2} \right]^{1/3} - 6.374 \times 10^6 \, m = 6.07 \times 10^7 \, m$$

What do you think? Is this altitude "small"?

Gravitational Potential Energy

In our previous look at gravitational potential energy, we treated only the approximate case near Earth's surface where the gravitational force on objects is very nearly constant. Here, we treat the more general case within the framework of the universal law of gravitation. For our system of two masses, M and m, separated by a distance r, the gravitational potential energy is a function of r and is given by

$$U(r) = -\frac{GMm}{r},$$

where it is assumed that $U = 0$ at $r = \infty$ is the chosen reference.

Now, the gravitational force obeys the **principle of superposition** which can be stated in the following way:

> *The net gravitational force on any given mass due to two or more other masses is the vector sum of the gravitational forces due to each of the other masses individually.*

A direct consequence of this fact for the potential energy is that gravitational potential energy is additive:

> *The total gravitational potential energy of a system of objects is the algebraic sum of the gravitational potential energies of each pair of objects.*

Having the gravitational potential energy, we can write down the total mechanical energy, $E = K + U$, of the system (Sun-planet or planet-satellite) as

$$E = \frac{1}{2}mv^2 - \frac{GMm}{r}.$$

This energy equation can be used to obtain a lot of information. One piece of information that can be determined from the total mechanical energy is the **escape speed** of a projectile. If a projectile is launched from the surface of the Earth such that the kinetic energy it is given exactly balances its potential energy, then $E = 0$ and the object will reach a distance $r = \infty$ where it will stop having both zero kinetic energy and zero potential energy. The speed that it was given at Earth's surface is called the escape speed

$$v_{esc} = \sqrt{\frac{2GM_E}{R_E}},$$

where M_E is the mass of the Earth and R_E is Earth's radius.

Types of Orbits

In addition to escape speed, the different types of orbits, that is, which of the conic sections an orbit will be, can be determined by the total mechanical energy. For elliptical orbits, the distance between a planet and the Sun varies. The closest approach ($r = r_{min}$) is called **perihelion** and the farthest distance ($r = r_{max}$) is called **aphelion**. The *semimajor axis* of the ellipse a is given by

$$a = \tfrac{1}{2}(r_{min} + r_{max});$$

and the *eccentricity* (or elongation) of the ellipse e is given by

$$e = (r_{max} - r_{min})/2a.$$

One can also define values of eccentricity for the other conic sections as well.

The conditions for the different types of orbits in terms of the total energy and eccentricity are summarized as follows:

Elliptical Orbits: $-\dfrac{GMm}{2a} < E < 0$ and $0 < e < 1$

Circular Orbits: $E = -\dfrac{GMm}{2a}$ and $e = 0$

Parabolic Orbits: $E = 0$ and $e = 1$

Hyperbolic Orbits: $E > 0$ and $e > 1$

Example 12–2 Space probes are often set in orbit around Earth before receiving a *boost* into space. If we have a probe in a circular orbit 3550 km above Earth's surface, how much of a kick in speed (a "Δv") does it need to achieve escape speed? (Assume that the boost is in the direction of orbital motion.)

Setting it up:

An object does not have to escape from the surface of the Earth. Replace the radius of the Earth with any greater radial distance from the center to determine the escape speed from that position. The information provided in the problem is the following:

Given: $h = 3550$ km; **Find:** Δv

Strategy:

We need to determine the speed that the probe has in the circular orbit and the escape speed at the position of the probe. Because the boost is in the same direction as the velocity in the circular orbit at the time of the boost, the Δv will be the difference between the two speeds.

Working it out:

In a circular orbit, gravity provides the centripetal force. So,

$$\frac{GM_E m}{r^2} = m\frac{v_{circ}^2}{r} \quad \Rightarrow \quad v_{circ} = \sqrt{\frac{GM_E}{r}}.$$

The escape speed is given by

$$v_{esc} = \sqrt{\frac{2GM_E}{r}}.$$

Therefore,

$$\Delta v = v_{esc} - v_{circ} = \sqrt{\frac{2GM_E}{r}} - \sqrt{\frac{GM_E}{r}} = \sqrt{\frac{GM_E}{r}}\left(\sqrt{2}-1\right).$$

Which gives

$$\Delta v = \sqrt{\frac{\left(6.67\times10^{-11}\,\frac{N\cdot m^2}{kg^2}\right)\left(5.976\times10^{24}\,kg\right)}{6.374\times10^6\,m + 3.55\times10^6\,m}}\left(\sqrt{2}-1\right) = 2.63\times10^3\,m/s.$$

What do you think? Does this result make it seem worthwhile to boost this probe from orbit?

Practice Quiz

1. Two objects are separated by a distance r. If the mass of each object is doubled, how does the gravitational force between them change?
   (a) It increases to twice as much.
   (b) It decreases to half as much.
   (c) It increases to four times as much.
   (d) It decreases to one-fourth as much.
   (e) None of the above

2. Two objects are separated by a distance r. If the distance between them is doubled, how does the gravitational force between them change?
   (a) It increases to twice as much.
   (b) It decreases to half as much.
   (c) It increases to four times as much.
   (d) It decreases to one-fourth as much.
   (e) None of the above

3. Which of the following statements is true according to Kepler's laws?
   (a) Planets orbit the Sun at constant speed.
   (b) Planets move slower when they are closer to the Sun.
   (c) Planets move slower when they are farther away from the Sun.
   (d) A planet is always the same distance from the Sun.
   (e) None of the above

4. A satellite orbits Earth at a distance r from Earth's center. If you double the distance, by approximately what factor will the period change?
   (a) 2.8   (b) 0.35   (c) 2.0   (d) 0.5   (e) 4.0

5. Two objects are separated by a distance r. If the mass of one object is reduced by half, how does the gravitational potential energy of the system change?
   (a) It increases, with U_f having half the magnitude of U_i.

(b) It decreases, with U_f having half the magnitude of U_i.

(c) It increases, with U_f having twice the magnitude of U_i.

(d) It decreases, with U_f having twice the magnitude of U_i.

(e) None of the above

6. Two objects are separated by a distance r. If the distance between them is tripled, how does the gravitational potential energy of the system change?

   (a) It increases, with U_f having three times the magnitude of U_i.

   (b) It decreases, with U_f having three times the magnitude of U_i.

   (c) It increases, with U_f having one-third the magnitude of U_i.

   (d) It decreases, with U_f having one-third the magnitude of U_i.

   (e) It decreases, with U_f having one-ninth the magnitude of U_i.

7. Assume that planet X has twice the mass and twice the radius of Earth. The escape speed from the surface of planet X, v_X, compared with that from the surface of Earth, v_E, is

   (a) $v_X = 2v_E$ **(b)** $v_X = \sqrt{2}v_E$ **(c)** $v_X = \frac{1}{2}v_E$ **(d)** $v_X = v_E/\sqrt{2}$ **(e)** $v_X = v_E$

12–4 Gravitation and Extended Objects

The law of universal gravitation as stated above applies only to point masses. We are able to apply this law to extended bodies, such as planets and stars, as a consequence of the principle of superposition. Particularly, when this principle is applied to extended bodies that are spherically symmetric the result is the following two statements (often called the shell theorem):

1. The net gravitational force experienced by a point mass located outside of a spherically symmetric object is identical to the force it would experience if the all of the mass of the spherical body was located at its center.

2. The net gravitational force experienced by a point mass located inside a spherically symmetric object is zero.

   Using the above theorem, we can see that the acceleration due to gravity at the surface of a spherically symmetric object, such as a planet, is given by

$$g = \frac{GM}{R^2},$$

where R is the radius of the object and M is its mass. As you travel above Earth's surface, you get further away from its center which affects the value of g that you experience. If we let h be your altitude (height above Earth's surface), then for small values of h, $h/R_E \ll 1$, we have

$$g_h = \frac{GM_E}{(R_E + h)^2} \approx \left[1 - \frac{2h}{R_E}\right] g_{\text{surface}},$$

when the binomial approximation is used.

Example 12–3 What is the acceleration due to gravity 100.0 ft above the surface of Mars?

Setting it up: The information given in the problem is the following:
Given: $h = 100.0$ ft;   **Find:** g

Strategy:
We need to adapt the expressions for the acceleration of gravity to Mars.

Working it out:
The acceleration at the surface of Mars is given by

$$g_{\text{surface}} = \frac{GM_M}{R_M^2}.$$

From the Appendix in your textbook, you can find the mass and radius of Mars.

Is an altitude of 100.0 ft small for the planet Mars?

$$\frac{h}{R_M} = \frac{100.0 \text{ ft}\left(\frac{0.3048 \text{ m}}{\text{ft}}\right)}{3{,}400{,}000 \text{ m}} = 8.94 \times 10^{-6} \ll 1;$$

so, the answer is yes. We can, therefore, use the approximation

$$g_h \approx \left[1 - \frac{2h}{R_M}\right]g_{\text{surface}} = \left[1 - 2(8.94 \times 10^{-6})\right]\frac{GM_M}{R_M^2}.$$

The final result is

$$g_h \approx \left[1 - 2(8.94 \times 10^{-6})\right]\frac{\left(6.67 \times 10^{-11} \text{ } \frac{\text{N·m}^2}{\text{kg}^2}\right)\left(0.11\right)\left(5.976 \times 10^{24} \text{ kg}\right)}{\left(3.4 \times 10^6 \text{ m}\right)^2} = 3.8 \text{ m/s}^2.$$

What do you think? Calculate the acceleration using the exact result. How good is this approximation?

Practice Quiz

8. What is the acceleration due to gravity at a height of 1000 km above Earth's surface?
   (a) 9.78 m/s^2 **(b)** 4.89 m/s^2 **(c)** 3.98 m/s^2 **(d)** 7.31 m/s^2 **(e)** None of the above

12–5—12-6 A Closer Look at Gravitation and Einstein's Theory of Gravity

You may notice that the universal law of gravitation does not distinguish between the masses involved. In this law, the masses m_1 and m_2 are completely interchangeable. Therefore, what is true for one mass is also true for the other. There *appears* to be a distinction, however, in that Earth orbits the Sun, but the Sun does not appear to orbit the Earth as the Moon does. In fact, the Sun does orbit the Earth. What happens is that both the Earth and the Sun are in orbit around the center of mass of the Sun-Earth system. (Actually, it's the center of mass of the solar system when all other bodies are included.) The Sun only appears to be stationary because its mass is so much greater than Earth's that this center of mass point is located very near the center of the Sun.

Another interesting fact about gravitation is that mass, an object's inertia, acts as a form of "gravitational charge" that determines the presence and strength of an object's gravitational interaction. There is no obvious physical reason why gravitational mass should be equivalent to inertial mass. However, the fact of this equivalence helped Einstein develop the theory of general relativity.

Reference Tools and Resources

I. Key Terms and Phrases

Newton's law of universal gravitation   between any two point masses there is an attractive force directly proportional to the product of the masses and inversely proportional to the square of the distance between them

Kepler's three laws of planetary motion   three laws that describe the behavior of orbiting bodies

principle of superposition   the net result of the gravitational interaction within a system of many particles is the sum of the results for interactions between each pair of particles in the system

escape speed   the speed at which a moving object can just barely escape the gravitational pull of another object to get infinitely far away

perihelion   the point at which a planet is closest to the Sun

aphelion   the point at which a planet is farthest from the Sun

II. Important Equations

Name/Topic	Equation	Explanation
Newton's law of universal gravitation	$\vec{F} = -\dfrac{GMm}{r^2}\hat{r}$	The vector form of the law of universal gravitation.
Kepler's third law of planetary motion	$\dfrac{T^2}{R^3} = \dfrac{4\pi^2}{GM}$	The mathematical expression of Kepler's third law.
Gravitational potential energy	$U(r) = -\dfrac{GMm}{r}$	The gravitational potential energy of two point masses assuming $U = 0$ when they are infinitely far apart.
Escape velocity	$v_{esc} = \sqrt{\dfrac{2GM_E}{R_E}}$	The speed required to escape from the surface of the Earth.
The acceleration of gravity	$g = \dfrac{GM}{R^2}$	The acceleration due to gravity a distance R from the center of a spherically symmetric body.

III. Know Your Units

Quantity	Dimension	SI Unit
Universal gravitation constant (G)	$[M^{-1}L^3T^{-2}]$	N·m²/kg²
Eccentricity (e)	dimensionless	—

Practice Problems

1. In the figure to the right, D is 2078 meters and d is 1725 meters. To the nearest MN, what is the magnitude of the gravitational force on the space ship? Consider all three to be point masses.

2. In the figure below, D is 2637 meters and d is 947 meters. To the nearest MN, what is the magnitude of the force on the space ship?

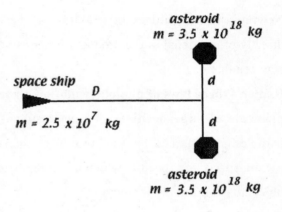

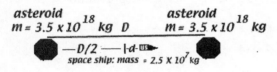

3. A planet/moon has a mass of 58.91×10^{23} kg and a radius of 11.18×10^6 m. To the nearest hundredth of a m/s², what is g on the surface?

4. In the figure the middle planet has a period of 1 year and an average distance from the Sun of 1.76 $\times 10^{11}$ meters. The average distance from the sun for the red planet is 0.55×10^{11} meters. To the nearest hundredth of a year, what is its period?

5. If the average distance of the rightmost planet in the above problem is 2.5×10^{11} meters, what is its period?

6. In the figure, D is 2.52×10^{11} meters and d is 1.05×10^{11} meters and the rightmost planet sweeps an area of 1×10^{14} m² in a given short unit of time while traveling at 4.72×10^4 m/s at D. To the nearest m/s, what is its speed at d?

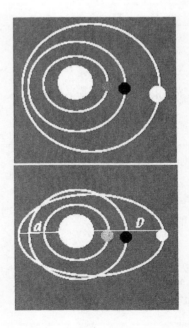

7. In the figure, m_1 = 2.56 kg, m_2 = 2.78 kg, and m_3 = 2.5 kg. To the nearest hundredth of a nJ, what is the potential energy of the configuration?

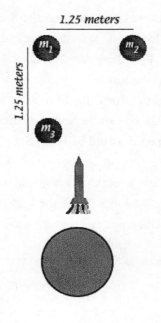

8. A planet has a mass of 8.28 × 10²⁴ kg and a radius of 5.8 × 10⁶ m. To the nearest m/s, what is its escape velocity?

9. In the previous problem, the mass is increased by a factor of 17. What is the new escape velocity?

10. In Problem 8, the radius is decreased by a factor of 0.49. What is its new escape velocity?

Selected Solutions to End of Chapter Problems

5. **(a)** The gravitational force between them is given by
$$F = GMm/r^2 = (6.67 \times 10^{-11} \text{ N·m}^2/\text{kg}^2)(95 \text{ kg})(68 \text{ kg})/(0.48 \text{ m})^2 = \boxed{1.9 \times 10^{-6} \text{ N.}}$$
(b) The attraction toward each other $\boxed{\text{must be the same}}$. Remember Newton's third law.

9. We find the radius from Kepler's third law:
$$T^2 = 4\pi^2 R^3/GM$$
which gives
$$R = \left(\frac{GMT^2}{4\pi^2}\right)^{1/3} = \left[\frac{\left(6.67 \times 10^{-11} \frac{\text{N·m}^2}{\text{kg}^2}\right)\left(5.98 \times 10^{24} \text{kg}\right)\left[90 \text{ min}\left(\frac{60 \text{ s}}{\text{min}}\right)\right]^2}{4\pi^2}\right]^{1/3} = \boxed{6.6 \times 10^6 \text{ m}}$$

25. We use conservation of energy, with the reference level for potential energy at infinity:
$$K_i + U_i = K_f + U_f$$
$$\tfrac{1}{2} mv_0^2 - GMm/D = \tfrac{1}{2} mv^2 - GMm/R$$
$$v^2 = v_0^2 + 2GM(1/R - 1/D)$$
$$v^2 = (2000 \text{ m/s})^2 + 2(6.67 \times 10^{-11} \text{ N·m}^2/\text{kg}^2)(5.98 \times 10^{24} \text{ kg})[1/(6.37 \times 10^6 \text{ m}) - 1/(8.00 \times 10^7 \text{ m})],$$
which gives $v = \boxed{1.09 \times 10^4 \text{ m/s}}$.

35. To escape from the solar system the escape speed needed, v_{esc}, satisfies
$$\tfrac{1}{2} Mv_{\text{esc}}^2 - GmM_E/R_E - GmM_S/r_{ES} = 0 \quad \Rightarrow \quad v_{\text{esc}} = [2G(M_E/R_E + M_S/r_{ES})]^{1/2}.$$
Plug in M_E = 5.98 × 10²⁴ kg, M_S = 1.99 × 10³⁰ kg, R_E = 6.38 × 10⁶ m, and r_{ES} = 1.50 × 10¹¹ m, to obtain v_{esc} = 43.5 km/s.

Earth moves once around the Sun in a near-circular orbital of radius $r = 1.5 \times 10^{11}$ m in t = one year = 3.156×10^7 s, so its (average) orbital speed is

$$V_E = 2\pi r_{ES}/t = 2\pi (1.5 \times 10^{11} \text{ m})/(3.156 \times 10^7 \text{ s}) = 2.99 \times 10^4 \text{ m/s} = 29.9 \text{ km/s},$$

relative to the Sun.

To escape from the Sun with a minimum possible speed v_{min}, launch the rocket in the same direction as $\vec{v}_E$, so the speed of the rocket relative to the Sun is $v = v_E + v_{min}$. (If $\vec{v}_E$ and $\vec{v}_{min}$ are not aligned then v would be less.) Set $v = v_{esc}$ to obtain

$$v_{min} = v_{esc} - v_E = 43.5 \text{ km/s} - 29.9 \text{ km/s} = 13.4 \text{ km/s} \approx \boxed{13 \text{ km/s}} .$$

(To be more accurate, one needs to consider the linear speed of Earth at the equator due to its spin. This is 0.464 km/s. So if the launch takes place at the equator we may further decrease v_{min} by that amount. But since r_{ES} is known to only 2 significant figures this does not affect the answer.)

47. The speed v of the object upon arriving at the surface is found from conservation of energy:

$$E_i \text{ (at infinity)} = E_f \text{(on the surface of the sphere)}$$

$$0 = \tfrac{1}{2} mv^2 - GMm/R \quad \Rightarrow \quad v = (2GM/R)^{1/2}.$$

Once it enters the hollow sphere, the gravitational acceleration of the object becomes zero (Newton's shell theorem), and the speed of the object does not change until it emerges from the other end of the sphere. Since the object has no angular momentum with respect to the sphere it must be aiming directly at the center of the sphere and therefore it travels a distance of $2R$ inside the sphere. The time it takes to travel through the sphere is then

$$t = 2R/v = 2R/(2GM/R)^{1/2} = \boxed{(2R^3/GM)^{1/2}}.$$

57. The gravitational force provides the centripetal acceleration:

$$GMM/(2R)^2 = MR\omega^2 \quad \Rightarrow \quad \omega^2 = GM/4R^3.$$

The period of the motion is $T = 2\pi/\omega = \boxed{4\pi (R^3/GM)^{1/2}}.$

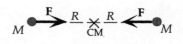

Answers to Practice Quiz
1. (c) 2. (d) 3. (c) 4. (a) 5. (a) 6. (c) 7. (e) 8. (d)

Answers to Practice Problems
1.	1,231 MN	2.	41,177 MN
3.	3.14 m/s²	4.	0.17 yr
5.	1.69 yr	6.	113,280 m/s
7.	–0.98 nJ	8.	13,803 m/s
9.	56,909 m/s	10.	19,718 m/s

CHAPTER 13

OSCILLATORY MOTION

Chapter Objectives

After studying this chapter, you should

1. understand the basic properties of oscillatory motion.
2. be able to describe simple harmonic motion mathematically.
3. be able to use the conservation of energy to analyze simple harmonic motion.
4. understand the behavior of simple and physical pendulums.
5. be able to identify the basic types of damped harmonic motion.
6. understand the phenomenon of resonance.

Chapter Review

In this chapter, we study what happens when objects in equilibrium are disturbed slightly away from this equilibrium. The most common result of such disturbances is that objects oscillate around the equilibrium configuration. The most important type of oscillatory motion is simple harmonic motion which is the main topic of this chapter.

13–1—13–2 The Kinematics of Simple Harmonic Motion and Circular Motion

Simple Harmonic Motion is the oscillatory motion that results from a Hooke's law force, a restoring force for which the magnitude is directly proportional to the displacement from equilibrium. The position of an object undergoing this type of motion can be described by the equations

$$x(t) = A\sin(\omega t + \delta)$$
$$= [A\cos\delta]\sin(\omega t) + [A\sin\delta]\cos(\omega t).$$

In this expression, A is the maximum displacement of the object from equilibrium called the amplitude, ω is called the **angular frequency** of the motion (explained below), and δ is the **phase**. The phase sets the initial position of the object at $t = 0$. Differentiating the position as a function of time, we obtain equations describing the velocity and acceleration of simple harmonic motion

$$v(t) = \omega A\cos(\omega t + \delta)$$
$$a(t) = -\omega^2 A\sin(\omega t + \delta) = -\omega^2 x(t).$$

The basic properties of this motion include the **period** T, which is the amount of time required for the motion to repeat itself; and the **frequency** f, which is the number of repetitions (or oscillations) per unit time. You may recall from circular motion that the period and the frequency are related by $f = 1/T$. Both of these quantities are related to the angular frequency in the above equation;

$$\omega = 2\pi f = \frac{2\pi}{T}.$$

So, these expressions show that if you think of an oscillation as taking up the 2π radians of a circular motion, then the angular frequency represents the rate, in radians per time, at which the motion takes place. In fact, this connection with circular motion can be very useful. Any simple harmonic motion can be viewed as the projection of uniform circular motion on the y-axis (or x-axis). Making this connection reveals that the angular frequency is identically equal to the constant angular velocity of the uniform circular motion.

Example 13–1 An object undergoes simple harmonic motion with a frequency of 3.50 Hz and an amplitude of 0.850 m. If its equilibrium position is $x = 0$ and its motion began at $x = A$, **(a)** what is the phase δ, **(b)** where is it 1.055 seconds later? **(c)** Write the complete equation of the motion.

Setting it up:
The information provided in the problem is the following:
Given: $f = 3.50$ Hz, $A = 0.850$ m, $x(0) = A$, $t_f = 1.055$ s;   **Find:** (a) δ, (b) $x(t_f)$, (c) $x(t)$

Strategy:
To solve this problem, we use the general equation for simple harmonic motion, $x(t) = A\sin(\omega t + \delta)$.

Working it out:
(a) Initially, at $t = 0$, we have $x(0) = A\sin(\delta) = A$. Therefore,

$$\delta = \sin^{-1}(1) = \pi/2 .$$

(b) To determine the location at $t = t_f$, we need to use the fact that $\omega = 2\pi f$. So, we can evaluate the position to be

$$x(t_f) = A\sin(2\pi f t_f + \delta) = [0.850\,\text{m}]\sin\left[2\pi(3.50\,\text{Hz})(1.055\text{ s}) + \tfrac{\pi}{2}\text{rad}\right] = -0.300\,\text{m} .$$

(c) The complete equation of the motion can be written as

$$x(t) = [0.850\,\text{m}]\sin\left[2\pi(3.50\,\text{Hz})t + \tfrac{\pi}{2}\text{rad}\right] .$$

What do you think?   What would the answers to parts (a), (b), and (c) be if the motion started at $x = 0$?

Practice Quiz

1. If the period of an oscillating body is 2.3 seconds, what is its frequency?
    (a) 2.3 Hz   **(b)** 0.43 Hz   **(c)** 2.3 s   **(d)** 0.43 s   **(e)** 2.7 Hz

2. An object oscillates with simple harmonic motion according to the equation
    $$x(t) = (0.15 \text{ m})\sin[(3.7 \text{ rad/s})t + 0.04 \text{ rad}] .$$
    What is the amplitude of this motion?
    (a) 0.15 m   **(b)** 3.7 m   **(c)** 1.7 m   **(d)** 0.94 m   **(e)** 0.59 m

3. An object oscillates with simple harmonic motion according to the equation
$$x(t) = (0.15 \text{ m}) \sin[(3.7 \text{ rad/s})t + 0.04 \text{ rad}].$$
What is the period of this motion?

(a) 0.15 s   **(b)** 3.7 s   **(c)** 1.7 s   **(d)** 0.94 s   **(e)** 0.59 s

13–3 Springs and Simple Harmonic Motion

As stated previously, simple harmonic motion is the motion that results from a Hooke's law force, this is the force exerted by an ideal spring, $F = -kx$. Combining this with Newton's second law we see that $a = -kx/m$. Since, for simple harmonic motion, $a = -\omega^2 x$, we can make the identification,

$$\omega = \sqrt{\frac{k}{m}}.$$

This equation leads to the further identifications

$$T = 2\pi \sqrt{\frac{m}{k}} \quad \text{and} \quad f = \frac{1}{2\pi} \sqrt{\frac{k}{m}}.$$

For cases in which a constant force, F_c, acts in conjunction with the restoring force, as with a mass hanging from a vertical spring, it turns out that the motion is still simple harmonic motion. The effect of the constant force is simply to shift the equilibrium position of the motion by an amount F_c/k.

Example 13–2 Consider an industrial spring of force constant 225 N/cm that supports an object of mass 5.03 kg. If the system is disturbed and begins to oscillate, find the period, frequency, and angular frequency of the oscillation.

Setting it up:

This is a vertical spring, so it has the constant force of gravity acting on it. The only effect of gravity is to shift the equilibrium position; it otherwise behaves as normal simple harmonic motion. The information given in the problem is the following:

Given: $k = 225$ N/cm, $m = 5.03$ kg;   **Find:** $T, f,$ and ω

Strategy:

The well-established expressions for these quantities can be used directly, even for a vertical spring.

Working it out:

Let us first convert the spring constant to SI units,

$$k = 225 \text{ N/cm} \left(\frac{100 \text{ cm}}{\text{m}} \right) = 22500 \text{ N/m}.$$

The period of the motion is

$$T = 2\pi \sqrt{\frac{m}{k}} = 2\pi \sqrt{\frac{5.03 \text{ kg}}{22500 \text{ N/m}}} = 0.0939 \text{s}.$$

The frequency of the motion is

$$f = \frac{1}{2\pi}\sqrt{\frac{k}{m}} = \frac{1}{2\pi}\sqrt{\frac{22500\,\text{N/m}}{5.03\,\text{kg}}} = 10.6\,\text{Hz}\,.$$

The angular frequency of the motion is

$$\omega = 2\pi f = \sqrt{\frac{k}{m}} = \sqrt{\frac{22500\,\text{N/m}}{5.03\,\text{kg}}} = 66.9\,\text{rad/s}\,.$$

What do you think? By what factor would each of these results change if k was doubled?

Practice Quiz

4. If the frequency, f, increases by a factor of 2, by what factor does the angular frequency increase?
    (a) 1/2   **(b)** 2π **(c)** 4   **(d)** $1/2\pi$ **(e)** 2

5. When the position of an object undergoing simple harmonic motion is at its maximum displacement from equilibrium, the speed of the object is
    (a) at its maximum
    (b) zero
    (c) half its maximum value
    (d) twice its maximum value
    (e) none of the above

6. If a mass of 2.6 kg oscillates with a period of 1.73 seconds on the end of a vertical spring, what is the force constant of the spring?
    (a) 4.5 N/m   **(b)** 34 N/m   **(c)** 7.7 N/m   **(d)** 0.029 N/m   **(e)** 1.5 N/m

13–4 Energy and Simple Harmonic Motion

Because a force obeying Hooke's law is conservative, the mechanical energy in simple harmonic motion is conserved

$$E = K + U = \tfrac{1}{2}mv^2 + \tfrac{1}{2}kx^2 = \text{constant}$$

The energy oscillates between kinetic and potential energy. The energy is all potential energy when the mass is at its amplitude ($x = A$), where $v = 0$. Therefore, the mechanical energy is given by

$$E = \tfrac{1}{2}kA^2$$

Similarly, the energy is all kinetic energy when the mass is at the equilibrium position ($x = 0$), where $v = v_{\text{max}}$ is the amplitude of the velocity oscillation; therefore, we can also write the mechanical energy as $E = \tfrac{1}{2}mv_{\text{max}}^2$. Since $v_{\text{max}} = A\omega$, we can write $E = \tfrac{1}{2}mA^2\omega^2$.

Example 13–3 A spring of force constant 142 N/m is hung vertically. An object of mass 1.25 kg is attached to the end of this spring. The mass is then pulled down an additional 8.37 cm from the equilibrium position and released. Use energy conservation to determine the object's speed when it is halfway between its central position and its amplitude.

Setting it up:

As in the previous Example, for a vertical spring, once the new equilibrium position is established, we can completely ignore gravity and treat the system as we would a horizontal spring. By stating that the mass is pulled a certain amount, then released, we are being told the amplitude of the motion. The information given in the problem is the following:

Given: $k = 142$ N/m, $m = 1.25$ kg, $A = 8.37$ cm, $x = A/2$;   **Find:** v

Strategy:

We must identify the most appropriate expression for the mechanical energy of the system, based on the given information, to solve this problem.

Working it out:

Knowing the amplitude of the motion, we can note that the total energy of the system is $E = \frac{1}{2}kA^2$. In general, at an arbitrary position, the total energy is given by $E = \frac{1}{2}mv^2 + \frac{1}{2}kx^2$. Setting these equal for the case at hand, $x = A/2$, we get

$$\tfrac{1}{2}kA^2 = \tfrac{1}{2}mv^2 + \tfrac{1}{2}k\left(\frac{A}{2}\right)^2 = \tfrac{1}{2}mv^2 + \tfrac{1}{8}kA^2.$$

Rearranging this gives

$$\tfrac{1}{2}mv^2 = \tfrac{1}{2}kA^2 - \tfrac{1}{8}kA^2 = \tfrac{3}{8}kA^2.$$

Solving this for the speed produces the equation

$$v = \sqrt{\frac{3k}{4m}}A = \sqrt{\frac{3(142 \text{ N/m})}{4(1.25 \text{ kg})}}(0.0837 \text{ m}) = 0.773 \text{ m/s}.$$

What do you think? Use a similar approach to find the position at which the speed is half its maximum.

Practice Quiz

7. When the kinetic energy of an object undergoing simple harmonic motion is maximum, its potential energy is
    (a) maximum
    (b) equal to the kinetic energy
    (c) half its maximum
    (d) twice the kinetic energy
    (e) none of the above

8. An object of mass 3.3 kg oscillates with SHM according to the equation

    $$x = (8.7 \text{ cm})\sin[(1.4 \text{ rad/s})t + 0.5 \text{ rad}].$$

    What is the mechanical energy of this system?
    (a) 29 J   (b) 40 J   (c) 3.6 J   (d) 0.024 J   (e) 0.0059 J

13–5—13–6 The Simple Pendulum and More

A **simple pendulum** is a point mass m suspended by a cord or rod (of negligible mass) of length ℓ. When a simple pendulum is displaced from equilibrium by small amounts, its oscillation about equilibrium is very nearly simple harmonic motion with gravity providing the restoring force. An analysis of this kind of system, using Newton's second law, shows that the "force constant" for a simple pendulum works out to be $k = mg / \ell$. We can use this result for k to characterize the motion of the simple pendulum in precisely the same way we used k previously. Therefore, substituting mg / ℓ for k, the period and the frequency of the motion are given by

$$T = 2\pi \sqrt{\frac{\ell}{g}} \quad \text{and} \quad f = \frac{1}{2\pi} \sqrt{\frac{g}{\ell}}.$$

Of course, the conservation of energy also applies to simple pendulums. If we take the lowest point as the reference level for zero gravitational potential energy, then the potential energy of the simple pendulum is given by

$$U = mgL(1 - \cos\theta) \approx \tfrac{1}{2} mg\ell\theta^2,$$

where θ is the angle, measured in radians, that the cord makes with the vertical.

For a simple pendulum, we ignore the mass of the rod that holds the bob. When more precision is needed, this approximation may not be good enough. A pendulum that does not meet the description of a simple pendulum is called a **physical pendulum**. For a physical pendulum, we must consider it to be a rotational system and take into account its moment of inertia, I. When we do this analysis, we find that the period of a physical pendulum of mass M is

$$T = 2\pi \sqrt{\frac{I}{Mgr}}$$

where r is the distance from the axis about which the pendulum swings to the center of mass of the system.

Example 13–4 Pendulums are sometimes used to measure the local acceleration of gravity. Mount Everest in Nepal is approximately 8850 m tall at its peak. If a simple pendulum of length 1.748 m set in small oscillations at the top of Mount Everest has a measured period of 2.655 s, what is the acceleration of gravity there?

Setting it up:
The information provided in the problem is the following:
Given: $h = 8850$ m, $\ell = 1.748$ m, $T = 2.655$ s;   **Find:** g

Strategy:
We know how the period relates to the acceleration of gravity for a simple pendulum; so we can use that expression to find the acceleration given that the period is known.

Working it out:

The equation for the period is

$$T = 2\pi\sqrt{\frac{\ell}{g}} \quad \Rightarrow \quad g = \frac{4\pi^2\ell}{T^2}.$$

Therefore,

$$g = \frac{4\pi^2\ell}{T^2} = \frac{4\pi^2(1.748\text{ m})}{(2.655\text{ s})^2} = 9.790\text{ m/s}^2.$$

What do you think? How does the local acceleration of gravity affect the frequency of the pendulum?

Practice Quiz

9. A meter stick is fixed to a horizontal rod at one end, while the other end is allowed to swing freely. A small Styrofoam ball is glued to the free-swinging end of the stick. This system should be treated as
    (a) a simple pendulum
    (b) a physical pendulum
    (c) none of the above

10. What is the period of small oscillations of the meter stick in question 9, if it has uniform mass?
    (a) 0.61 s   (b) 3.83 s   (c) 2.00 s   (d) 0.26 s   (e) 1.64 s

13–7—13–8 Damped and Driven Harmonic Motion

In realistic oscillating systems, energy is lost during the motion. Because of this energy loss, the amplitude of the oscillation decreases. We call this type of motion *damped oscillation*. Frequently, damping forces (like air resistance) are proportional to the velocity of the object on which the force is applied, so that

$$\vec{F} = -b\vec{v},$$

where the coefficient b is called the *damping coefficient*. For motion along a single dimension x, an analysis of oscillatory motion, with the above damping force, shows that the motion is given by

$$x(t) = Ae^{-\alpha t}\sin(\omega't + \delta)$$

where, as with simple harmonic motion, A is the maximum amplitude of the oscillation and δ is the phase. The damping factor α and the angular frequency ω' are given by

$$\alpha = \frac{b}{2m} \quad \text{and} \quad \omega' = \sqrt{\omega_0^2 - \frac{b^2}{4m^2}},$$

where ω_0 is called the **natural frequency** of the system; it is the angular frequency the system would have were it not for the damping force.

From the equation for ω' you can see that at a special value of the damping coefficient, $b_c = 2m\omega_0$, the angular frequency is zero. When $b = b_c$, the system is just at the condition when oscillations no longer occur; we call this **critically damped** motion. When $b < b_c$, the system clearly undergoes some oscillations as the motion dies out; we call this **underdamped** motion. When $b > b_c$, the motion of the system dies out with no sign of oscillation; we call this **overdamped** motion.

How quickly a damped oscillation dies out is characterized by two quantities. The *lifetime* τ is defined to be $\tau = m/b$ and represents an amount of time for the amplitude of the oscillation to decrease most of the way to zero. Also, a dimensionless quantity called the *Q factor*, given by $Q = \omega_0 \tau$, also characterizes this damping time.

The effects of damping forces can be overcome by putting energy into the system in order to maintain the oscillation. When this energy is input we say that the motion is a *driven* oscillation. When a damped system is being driven to oscillate at a frequency ω by a driving force $F = F_0 \sin(\omega t)$, the amplitude is given by

$$A = \frac{F_0}{\sqrt{m^2 \left(\omega^2 - \omega_0^2\right)^2 + b^2 \omega^2}} .$$

The above equation shows that the amplitude of the motion increases as the frequency of the driving force approaches the natural oscillation frequency of system. This phenomenon is called **resonance** (for this reason natural frequencies are sometimes called *resonant frequencies*). Resonance plays an important role in many applications including the generation and detection of sound to be studied later.

Example 13–5 A child that weighs 45 lb enjoys swinging very high. If the length of the chain that holds the swing is 2.5 m, approximately how often should the child be pushed to get a good high swing?

Setting it up:

Asking "how often" the child should be pushed is another way of asking for the frequency. Since we want the push to produce "a good high swing" we are looking for a large amplitude of oscillation. This fact means that the pushes should be at resonance. The information provided is the following:
Given: $W = 45$ lb, $\ell = 2.5$ m ;   **Find:** f

Strategy:

Since we seek only an approximate answer, we will make two approximations. We will treat the child and swing as a simple pendulum with the child as the bob, and we will use the simple harmonic motion results and ignore any slight inaccuracies that result from the fact that the child swings through large angles.

Working it out:

The natural frequency for the child is approximately,

$$f = \frac{1}{T} = \frac{1}{2\pi} \sqrt{\frac{g}{\ell}} = \frac{1}{2\pi} \sqrt{\frac{9.80 \text{ m/s}^2}{2.5 \text{ m}}} = 0.32 \text{ Hz} .$$

Because $0.32 \approx 1/3$, we can say that there should be approximately 1 push every 3 seconds.

What do you think? Do the assumptions made suggest that this approximation is a little too high or a little too low?

Practice Quiz

11. If the child in the swing in Example 13–5 experiences a damping force with a damping coefficient of 4.82 kg/s, what is the lifetime of the damped oscillation?

    (a) 0.24 s   (b) 4.2 s   (c) 20.4 s   (d) 4.6 s   (e) 8.4 s

Reference Tools and Resources

I. Key Terms and Phrases

simple harmonic motion the oscillatory motion that results from a force that obeys Hooke's law

frequency the number of cycles per unit of time

angular frequency 2π times the frequency

phase an angle used to indicate the starting point of an oscillation

amplitude the maximum displacement from equilibrium

period the amount of time for one complete cycle

simple pendulum a mass suspended by a cord or rod of negligible mass

physical pendulum a mass distribution that is suspended and free to oscillate

natural frequency a frequency at which a system will naturally oscillate about stable equilibrium in the absence of drag

underdamped when a small damping constant causes an oscillating system to decrease in amplitude

critically damped when the damping constant is just large enough to prevent oscillations

overdamped when the damping constant is more than enough to prevent oscillations

resonance a phenomenon in which large-amplitude oscillations occur when a system is driven at its natural frequency

II. Important Equations

Name/Topic	Equation	Explanation
Simple harmonic motion	$x(t) = A\sin(\omega t + \delta)$	The displacement as a function of time for simple harmonic motion.
Frequency	$f = 1/T;\ \omega = 2\pi f$	The frequency and angular frequency of simple harmonic motion.
Period	$T = 2\pi\sqrt{m/k}$	The period of the simple harmonic motion of a mass, m, oscillating on a spring of spring constant k.
Mechanical energy	$E = \frac{1}{2}kA^2 = \frac{1}{2}mA^2\omega^2$	The total mechanical energy of an object in simple harmonic motion.

Pendulums	$T = 2\pi\sqrt{\ell/g}$ $$T = 2\pi\sqrt{I/Mgr}$$	The period for a simple pendulum and a physical pendulum, respectively.
Damped harmonic motion	$\omega' = \sqrt{\omega_0^2 - \dfrac{b^2}{4m^2}}$	The frequency of an object undergoing damped harmonic motion.
Resonance	$A = \dfrac{F_0}{\sqrt{m^2\left(\omega^2 - \omega_0^2\right) + b^2\omega^2}}$	How the amplitude of oscillation depends on the relationship between the driving frequency and the natural frequency.

III. Know Your Units

Quantity	Dimension	SI Unit
Damping coefficient (b)	$[MT^{-1}]$	kg/s

Practice Problems

1. The velocity of a simple harmonic oscillator is given by $v = -(4.58)\sin(21.9t)$, in mks units. What is its angular frequency?

2. What is the amplitude of the motion in Problem 1, in meters, to two decimal places?

3. To the nearest hundredth of a meter, where is the mass in Problem 1 at the time $t = 2.68$ seconds?

4. The mass in Problem 3 is 0.71 kg. To the nearest tenth of a joule, what is the spring's potential energy?

5. What is its kinetic energy to the nearest tenth of a joule?

6. A 0.37-kg mass is attached to a spring and let fall. The spring constant is 5.6 N/m. To the nearest hundredth of a meter, what is the point where it "stops"?

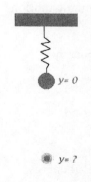

7. In Problem 6, what is the amplitude of the resulting motion?

8. A 28-g bullet traveling at 160 m/s is fired into a 0.422-kg wooden block anchored to a 100 N/m spring. How far is the spring compressed, to the nearest thousandth of a meter?

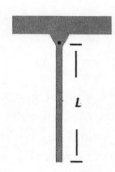

9. The speed of the bullet in Problem 8 is not known, but it is observed that the spring is compressed 48.6 cm. To the nearest m/s, what was the speed of the bullet?

10. A wooden rod of mass $m = 0.45$ kg and length $L = 0.7$ m is used as a physical pendulum. To the nearest tenth of a second, what is the period of oscillation?

Selected Solutions to End of Chapter Problems

5. The period is related to the angular velocity, which we can get from $v_{max} = A\omega$,

$$\omega = v_{max}/A = (4.36 \text{ m/s})/2.84 \text{ m} = 1.54 \text{ rad/s.}$$

The period is then

$$T = 2\pi/\omega = 2\pi/(1.54 \text{ rad/s}) = \boxed{4.09 \text{ s}}.$$

13. (a) We find the angular frequency from

$$\omega = \sqrt{k/m} = \sqrt{(0.50 \text{ N/m})/(0.20 \text{ kg})} = 1.58 \text{ rad/s.}$$

The period is

$$T = 2\pi/\omega = 2\pi/(1.58 \text{ rad/s}) = \boxed{4.0 \text{ s}}.$$

(b) From $v_{max} = A\omega$, we get

$$A = v_{max}/\omega = (2.0 \text{ m/s})/(1.58 \text{ rad/s}) = \boxed{1.3 \text{ m}}.$$

29. The period depends only on the mass and force constant:

$$T = 2\pi\sqrt{m/k} \quad \Rightarrow \quad k = \frac{4\pi^2 m}{T^2} = \frac{4\pi^2 (0.045 \text{ kg})}{(3.1 \text{ s})^2} = \boxed{0.18 \text{ N/m}}$$

45. The angular frequency is given by

$$\omega = \sqrt{k/m} = \sqrt{(1 \text{ N/m})/(0.2 \text{ kg})} = 2.24 \text{ rad/s}$$

Because $x = x_{max}$ at $t = 0$, we can write

$$x = x_{max} \sin(\omega t + \pi/2).$$

Taking the derivative of this we get

$$v = x_{max}\omega \cos\left(\omega t + \pi/2\right) = v_{max}\cos\left(\omega t + \pi/2\right)$$

We can use the data at $t = 0.5$ s to get the maximum speed from

$$v_{max} = \frac{v}{\cos\left(\omega t + \pi/2\right)} = \frac{1.5 \text{ m/s}}{\cos\left[\left(2.24\frac{rad}{s}\right)\left(0.5 \text{ s}\right) + \pi/2\right]} = \boxed{2 \text{ m/s}}$$

With the above result, we can now say

$$x_{max} = v_{max}/\omega = (1.67 \text{ m/s})/(2.24 \text{ rad/s}) = \boxed{0.7 \text{ m}}.$$

The total energy is

$$E = \tfrac{1}{2}kx_{max}^2 = \tfrac{1}{2}(1 \text{ N/m})(0.74 \text{ m})^2 = \boxed{0.3 \text{ J}}.$$

57. Set the zero level of gravitational potential energy at the lowest point of the pendulum bob's swing. Then as the pendulum swings to the maximum angle α the bob's vertical displacement is $h = L(1 - \cos\alpha)$, whereupon the potential energy of the bob is

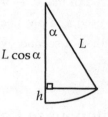

$$U = mgh = mgL\,(1 - \cos\alpha)$$

Note that, at that point, the bob is not moving and so $K = K_{min} = 0$ and

$$U = U_{max} = E \quad \Rightarrow \quad E = \boxed{mgL(1 - \cos\alpha)}.$$

If the angle α is small, then the approximation $\cos\alpha \approx 1 - \tfrac{1}{2}\alpha^2$ may be applied, which yields

$$E \approx mgL\left[1 - \left(1 - \tfrac{1}{2}\alpha^2\right)\right] = \boxed{\tfrac{1}{2}mgL\alpha^2} \qquad \left(|\alpha| \ll 1\right)$$

79. We find the effective force constant from the equilibrium condition,

$$k\,\Delta y = mg \quad \Rightarrow \quad k/m = g/\Delta y$$

Resonance will occur if the driving frequency equals the natural frequency,

$$\omega_0 = \sqrt{k/m} = \sqrt{g/\Delta y} = \sqrt{\left(9.80\tfrac{m}{s^2}\right)/\left(8\times10^{-2}\text{m}\right)} = 11 \text{ rad/s}$$

Therefore,

$$f = \omega_0/2\pi = 11\tfrac{rad}{s}/2\pi = \boxed{1.8 \text{ Hz}}$$

Answers to Practice Quiz

1. (b) **2.** (a) **3.** (c) **4.** (e) **5.** (b) **6.** (b) **7.** (e) **8.** (d) **9.** (b) **10.** (e) **11.** (b)

Answers to Practice Problems

1. 21.9 rad/s   **2.**   0.21 m

3. −0.11 m   **4.**   2.0 J

5. 5.4 J   **6.**   1.30 m

7. 0.65 m   **8.**   0.668 m

9. 116 m/s   **10.** 1.4 s

CHAPTER 14

WAVES

Chapter Objectives

After studying this chapter, you should

1. know the difference between transverse, longitudinal, and standing waves.
2. understand the basic properties of periodic waves.
3. calculate the approximate speeds of a sound wave in air, a sound wave along a rod, and a shear wave in a solid.
4. be able to calculate the power delivered by waves.
5. be able to determine the allowed wavelengths and frequencies of standing waves on strings and in pipes.
6. be able to calculate the loudness of a sound on the decibel scale.
7. be able to determine frequency shifts due to the Doppler effect.

Chapter Review

In this chapter, we study waves. You can view a wave as resulting from the connection of a series of oscillators (oscillations were studied in the previous chapter) or as a propagating oscillation. Generally, any disturbance that propagates can be called a wave. The study of waves is important in almost every branch of physics and has many applications.

14–1—14–3 Types of Waves, the Wave Equation, and Periodic Waves

There are two main types of waves. These types are distinguished by the relationship between the direction of the oscillation of the medium in which the wave is traveling and the direction of propagation of the wave. In a **transverse wave**, the direction of oscillation is perpendicular (or transverse) to the direction of propagation. A wave on a string is a good example of a transverse wave. The other main type of wave is a **longitudinal wave**, in which the direction of oscillation is along the same line as the direction of propagation. A compression wave traveling along a spring (such as a slinky) is a good example of a longitudinal wave.

An important result of wave motion is the existence of **standing waves**. A standing wave is a regular wave pattern resulting from harmonic displacements taking place within the medium. In a particular medium, standing waves can only exist with certain frequencies. This fact explains why a certain string in a piano has a certain pitch.

The most common type of waves are periodic waves for which standing waves are an example. The main characteristics of periodic waves are related to the cycle of this repeating motion. One of these characteristics relates to the minimum time it takes for a wave to repeat itself, the *period*, τ. As with any simple harmonic motion, the inverse of the period is called the *frequency*, f. The frequency is closely related to the angular frequency $\omega = 2\pi f$.

Waves also repeat spatially; the minimum repeat length of a wave is called its **wavelength**, λ. If you consider the following diagram of a transverse wave, the wavelength equals the distance between successive crests, or troughs, of the wave (other corresponding points may also be used).

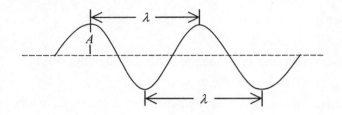

The wavelength is closely associated with a quantity called the **wave number**

$$k = 2\pi/\lambda,$$

which is a measure of the number of waves (in units of 2π radians per wave) per unit length. As the diagram shows, the amplitude, A, is the maximum displacement of the medium from equilibrium.

Another important characteristic of a wave is its speed of travel. This speed equals the distance the wave travels before it repeats (the wavelength) divided by the time it takes the wave to repeat (the period); therefore, the speed of a wave is given by

$$v = \frac{\lambda}{\tau} = \frac{\omega}{k} = \lambda f \ .$$

All of the above properties can be developed from a dynamical equation that governs the behavior of periodic waves, called the wave equation. Using a transverse wave on a string as a prototype, the wave equation is

$$\frac{\partial^2 z(x,t)}{\partial x^2} = \frac{1}{v^2} \frac{\partial^2 z(x,t)}{\partial t^2},$$

where z is the direction of oscillation of the string, x is the direction of propagation of the wave, and v is the speed of propagation. For a traverse wave on a string, this speed is

$$v = \sqrt{\frac{T}{\mu}},$$

where T is the tension in the string and μ is its mass per unit length.

Example 14–1 A string of length 2.59 m has a mass of 5.11 g and is fixed at one end. A person takes the other end and oscillates it up and down with a frequency of 3.47 Hz. If it takes the resulting wave 0.862 s to travel the length of the string, **(a)** determine the tension in the string, and **(b)** determine the wavelength of the wave.

Setting it up:
The information provided in the problem is the following:
Given: $L = 2.59$ m, $m = 5.11$ g, $f = 3.47$ Hz, $t = 0.862$ s;   **Find: (a)** τ, **(b)** λ

Strategy:

The given information allows us to determine both v and μ so that we can use the relationship between v, μ, and τ to solve part (a). Knowing v and frequency, we can then calculate the wavelength in part (b).

Working it out:

(a) The relationship between v, μ, and τ is

$$v = \sqrt{\frac{T}{\mu}} \quad \Rightarrow \quad T = \mu v^2,$$

where $\mu = m/L$ and $v = L/t$. This gives,

$$T = \left(\frac{m}{L}\right)\left(\frac{L}{t}\right)^2 = \frac{mL}{t^2} = \frac{(0.00511 \text{ kg})(2.59 \text{ m})}{(0.862 \text{ s})^2} = 0.0178 \text{ N}.$$

(b) The relationship between v, f, and λ is

$$v = \lambda f \quad \Rightarrow \quad \lambda = v/f.$$

Once again using $v = L/t$, we get

$$\lambda = \frac{L}{tf} = \frac{2.59 \text{ m}}{(3.47 \text{ Hz})(0.862 \text{ s})} = 0.866 \text{ m}.$$

What do you think? We never needed the values of the speed or the linear density, what are they?

Practice Quiz

1. The distance between two troughs of a wave of frequency 200 Hz is 1.7 m. What is the speed of this wave?

   (a) 5.0 mm/s   (b) 340 m/s   (c) 8.5 mm/s   (d) 120 m/s   (e) none of the above

2. If two waves traveling through the same medium have the same period, but wave 1 has half the wavelength of wave 2, how do the speeds of these waves compare?

   (a) $v_1 = 2v_2$ (b) $v_1 = 4v_2$ (c) $v_2 = 4v_1$ (d) $v_2 = 2v_1$ (e) $v_2 = v_1$

3. A certain string sustains waves of speed v when under a tension T. If the tension in the string is reduced to $T' = \frac{1}{3}T$, is the speed of the wave reduced or enhanced, and by what factor is it reduced or enhanced?

   (a) reduced by a factor of 0.58
   (b) reduced by a factor of 3
   (c) enhanced by a factor of 1.7
   (d) enhanced by a factor of 3
   (e) the speed stays the same

14–4—14–5 Traveling Waves, Energy, & Power

Solutions to the wave equation for a string represent disturbances that travel along the string, called traveling waves. The solution for a wave traveling in the positive x direction is

$$z(x,t) = A\sin(kx - \omega t) = A\sin[k(x - vt)],$$

and for a wave traveling in the negative x direction,

$$z(x,t) = A\sin(kx + \omega t) = A\sin[k(x + vt)].$$

For longitudinal waves, or **sound** waves, along a rod, we can show that the motion of the wave obeys the same wave equation except that the speed of the wave is given by

$$v = \sqrt{\frac{Y}{\rho}},$$

where Y is Young's modulus and ρ is the density of the medium. Noting the similarity between this expression and the one for the speed of a wave on a string, we can see that the speed of a wave propagating through a medium depends on the properties of the medium such that

$$\text{wave speed} = \sqrt{\frac{\text{restoring force factor}}{\text{mass factor}}}.$$

Using this fact, we can see the plausibility of how the speed of other types of waves depend on the medium in which the wave propagates. For example, shear waves in solids propagate at a speed given by

$$v_{\text{shear}} = \sqrt{\frac{G}{\rho}},$$

where G is the shear modulus. The speed of sound in air is given by

$$v_{\text{sound}} = \sqrt{\frac{\gamma p_0}{\rho_0}},$$

where p_0 and ρ_0 are the ambient pressure and density of the air, respectively, and $\gamma \cong 1.4$ (is a dimension factor that relates to the thermal properties of the gas). A typical value for the speed of sound in air is approximately 330 m/s.

The propagation of mechanical waves involves the motion of the medium; therefore, it is not surprising that waves deliver energy and power. A general characteristic of waves is that the energy and power carried by waves depend on the square of the amplitude and frequency of the waves. The average power is given by

$$\langle P \rangle = \tfrac{1}{2} \mu \omega^2 A^2 v,$$

and the energy density (energy per unit volume), u, carried along with the wave is

$$u(x,t) = \mu \omega^2 A^2 \cos^2(kx - \omega t).$$

Example 14–2 A transverse traveling wave is described by the function

$$z(x,t) = (3.2 \text{ m})\sin\left[\left(2.5 \text{ m}^{-1}\right)x - \left(1.7 \text{ s}^{-1}\right)t\right].$$

What are the **(a)** frequency, and **(b)** wavelength of this wave?

Setting it up:

We can compare the given function to the general solution for such a wave, $z(x,t) = A\sin(kx - \omega t)$. Doing the comparison shows that the information given in the problem is the following:

Given: $A = 3.2$ m, $k = 2.5$ rad/m, $\omega = 1.7$ rad/s;   **Find: (a)** f, **(b)** λ

Strategy:

Now that the given quantities have been properly identified, we can get the frequency from the angular frequency and the wavelength from the wave number.

Working it out:

(a) The relationship between the frequency and the angular frequency is $\omega = 2\pi f$. Thus,

$$f = \frac{\omega}{2\pi} = \frac{1.7 \text{ rad/s}}{2\pi \text{ rad}} = 0.27 \text{ Hz}.$$

(b) The relationship between the wavelength and the wave number is $k = 2\pi/\lambda$. Thus,

$$\lambda = \frac{2\pi}{k} = \frac{2\pi \text{ rad}}{2.5 \text{ rad/m}} = 2.5 \text{ m}.$$

What do you think?   What are the speed and period of this wave?

Practice Quiz

4.   What is the speed of a wave described by the expression $z = (0.21 \text{ m})\sin\left[(0.13 \text{ m}^{-1})x - (2.4 \text{ s}^{-1})t\right]$?

   (a) 48 m/s   **(b)** 0.38 m/s   **(c)** 18 m/s   **(d)** 0.21 m/s   **(e)** 2.4 m/s

14–6—14–7 Standing Waves and More About Sound

For the case of a wave on a string with fixed ends, an important solution of the wave equation is the following;

$$z(x,t) = z_0 \cos(\omega t + \phi)\sin kx,$$

where z_0 and ϕ are two arbitrary constants. The above solution is a standing wave solution. A standing wave oscillates in time but is fixed in space. Standing waves contain positions (or regions), called *nodes*,

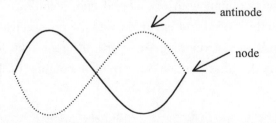

where no oscillation occurs and other positions where the oscillation of the medium has maximum amplitude, called *antinodes*. The condition that the ends remain fixed can be used to see that many standing waves of different wavelengths, called **modes**, can be established. For a string of length L, these allowed wavelengths are given by

$$\lambda_n = \frac{2}{n} L \quad \text{for } n = 1, 2, 3, \dots .$$

The mode corresponding to $n = 1$ is called the *fundamental mode*, and sometimes called the *first harmonic*, the modes for $n = 2$ and up are called the *second mode* and so on.

Each standing wave mode has a corresponding frequency. When working in terms of frequency instead of wavelength, we think in terms of the **harmonics** of the system. The harmonics of the system are the different frequencies of the allowed modes. Given the fact that

$$v = \lambda_n f_n = \sqrt{T/\mu}$$

we can determine that the harmonics are

$$f_n = \frac{n}{2L} \sqrt{\frac{T}{\mu}} \quad \text{for } n = 1, 2, 3 \dots$$

which also correspond to

$$\omega_n = \frac{n\pi}{L} \sqrt{\frac{T}{\mu}}$$

$$\tau_n = \frac{2L}{n} \sqrt{\frac{\mu}{T}}.$$

The frequency for $n = 1$ is called the **fundamental frequency**; and for $n = 2$ and up are called the *second harmonic* and so on.

Standing waves in air columns are understood in ways very similar to those for standing waves on strings; however, it is important to recognize that in air columns we are talking about longitudinal sound waves rather than transverse vibrations. Two types of air columns are in common use for generating sounds: one type has one end of the column closed off (with the other end open), and the other type has both ends open.

Standing waves in an air column of length L, with both ends open, display the same characteristics as waves on a string with both ends fixed:

$$\lambda_n = \frac{2}{n} L \quad \text{and} \quad f_n = \frac{n}{2L} v_{\text{sound}}$$

for $n = 1, 2, 3, \dots$. In an air column with one closed end, standing waves exhibit somewhat different characteristics. In this case the closed end is a node, and the open end is an antinode. Thus, unlike in the two previous cases the ends of the standing wave are different. The longest wavelength of a standing wave (the fundamental mode) that can be set up in a column of length L is $\lambda_1 = 4L$. The other wavelengths are

$$\lambda_n = \frac{4L}{n} \quad \text{with} \quad n = 1, 3, 5, \dots$$

For the frequencies, only odd harmonics (odd multiples of the fundamental frequency) are present:

$$f_n = n\frac{v_{\text{sound}}}{4L} \quad \text{with} \quad n = 1, 3, 5, \ldots$$

All the above information can be used to explain the generation of sound, especially music.

Example 14–3 A certain guitar string of length 0.90 m has a linear density of 0.0075 kg/m. When properly tuned, the string has a 4th harmonic of 1024 Hz. What is the tension for properly tuning this string?

Setting it up: A sketch of the 4th harmonic is helpful.
The information given in the problem is the following:
Given: $L = 0.90$ m, $\mu = 0.0075$ kg/m, $f_4 = 1024$ Hz
Find: T

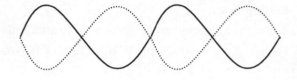

Strategy:
The tension in the string is related to the frequency of a given harmonic in a known way.

Working it out:
The relationship between tension and frequency is

$$f_n = \frac{n}{2L}\sqrt{\frac{T}{\mu}} \quad \Rightarrow \quad T = \frac{4\mu L^2}{n^2}f_4^2.$$

Therefore, we get

$$T = \frac{4\mu L^2}{n^2}f_4^2 = \frac{4(0.0075 \text{ kg/m})(0.90 \text{ m})^2}{4^2}(1024 \text{ Hz})^2 = 1600 \text{ N}.$$

What do you think? What are the wavelength, period, and speed of this wave?

Hearing Sounds

Frequency is the physical property of a sound wave that determines the human perception of a sound's pitch. The human perception of the loudness of a sound is determined by the **intensity**, I, of the wave carrying the sound. The average intensity of a wave is the amount of energy that passes through a given area per unit time divided by the area, that is, energy per unit time per unit area. Since energy per unit time is power (in watts) delivered by the wave, on average the intensity is given by $I = P/A$, where P is the power, and A is the area over which this power is spread. As can be seen from this relation, the SI unit of intensity is W/m^2.

The range of intensities that humans hear is very large, a more convenient measure, sometimes called the *intensity level*, β, measures loudness by comparing the intensity of a sound to a standard reference intensity, $I_0 = 10^{-12}$ W/m^2; this ratio is then rescaled by taking its logarithm. This quantity β is dimensionless; values of β are quoted on the **decibel scale** (dB). In mathematical terms, values of β can be obtained from the intensity I by the expression

$$\beta = 10\log\left(\frac{I}{I_0}\right).$$

When more than one source of sound contributes, the intensities of the individual sources add and β is determined from the resulting sum.

Example 14–4 Given that the lower threshold intensity for human hearing is about 1.0×10^{-12} W/m^2 and the pain threshold is about 1.0 W/m^2, determine the desirable range of sound in decibels.

Setting it up:

The information given in the problem is the following:

Given: $I_{lower} = 1.0 \times 10^{-12}$ W/m^2, $I_{upper} = 1.0$ W/m^2;   **Find:** β_{lower} and β_{upper}

Strategy:

Here we make direct use of the expression for calculating β from intensity.

Working it out:

The lower threshold in decibels corresponds to the reference intensity $I = I_0$. Therefore,

$$\beta_{lower} = 10\log\left(\frac{I}{I_0}\right) = 10\log\left(\frac{I_0}{I_0}\right) = 10\log(1) = 0 \text{ dB}$$

The upper part of the desirable range is

$$\beta_{upper} = 10\log\left(\frac{I}{I_0}\right) = 10\log\left(\frac{1.0 \text{ W/m}^2}{1.0 \times 10^{-12} \text{ W/m}^2}\right) = 10\log(10^{12}) = 120\log(10) = 120 \text{ dB}$$

Notice here that by using intensity level β instead of intensities I, the range of values you have to work with is much smaller without any loss of information

What do you think? How many orders of magnitude does the range include in terms of intensities and decibels.

Practice Quiz

5. What is the wavelength of the third harmonic for standing waves on a string of length 1.35 m with fixed ends?
    (a) 1.5 m   **(b)** 1.35 m   **(c)** 0.900 m   **(d)** 4.05 m   **(e)** 0.45 m

6. What is the wavelength of the third harmonic for standing waves in an air column of length 1.35 m with only one open end?
    (a) 5.4 m   **(b)** 1.35 m   **(c)** 1.80 m   **(d)** 1.08 m   **(e)** 0.45 m

7. If the intensity of a sound becomes a factor of 10 greater, describe what happens to the decibel level.
    (a) The decibel level increases by 1 dB.

(b) The decibel level decreases by 1 dB.

(c) The decibel level increases by a factor of 10.

(d) The decibel level increases by a factor of log(10).

(e) None of the above

14–8 The Doppler Effect

When there is relative motion between the source of a sound and the receiver of the sound, the pitch of the sound changes. The pitch gets higher if the source and the receiver are moving closer to each other, and it gets lower if they are moving further apart. The phenomenon just described is called the **Doppler effect**. This effect has analogs for all other types of waves in addition to sound.

Because the pitch of a sound is directly associated with its frequency, we treat the Doppler effect by finding the frequency perceived by the receiver, f', as compared with the frequency emitted by the source f_0. We take the speed of sound in air to be constant, $v = 330$ m/s, and call v_r the speed of the receiver and v_s the speed of the source. With these definitions, and the provision that both v_r and v_s be less than v, we represent the Doppler effect for sound by the equation

$$f' = \left(\frac{v - v_r}{v - v_s}\right) f_0 .$$

When using the above equation, we take the direction from the source toward the receiver as the positive direction; so, $v > 0$. If the receiver is moving in the direction away from the source, then $v_r > 0$, if toward the source $v_r < 0$. If the source is moving in the direction away from the receiver, then $v_s < 0$, if toward the receiver then $v_s > 0$. Thus, there are four different possible circumstances.

Example 14–5 **(a)** A car moves with a speed of 45.0 mi/h toward a stationary listener as its horn is blown, emitting a frequency of 445 Hz. What frequency is heard by the listener? **(b)** Two cars move toward each other, each with a speed of 22.5 mi/h. If one car's horn is blown with a frequency of 445 Hz, what frequency is heard by observers in the other car?

Setting it up:

We need to properly identify the velocities of sound, the source, the receiver, and their algebraic signs. If we take the direction from source to receiver as the positive direction, then the velocity of the source is positive. Therefore, the information provided in the problem can be written the following way:

Given: $v = 330$ m/s, $f_0 = 445$ Hz; **(a)** $v_s = 45.0$ mi/h, $v_r = 0$, **(b)** $v_s = 22.5$ mi/h, $v_r = -22.5$ mi/h;
Find: (a) f', **(b)** f'

Strategy:

With all relevant quantities identified, we only need to apply our equations for the Doppler effect.

Working it out:

(a) Let's first convert the source velocity to SI units;

$$v_s = 45.0 \text{ mi/h}\left(\frac{0.447 \text{ m/s}}{\text{mi/h}}\right) = 20.115 \text{ m/s}$$

The expression for the Doppler effect gives

$$f' = \left(\frac{v - v_r}{v - v_s}\right)f_0 = \left(\frac{330 \text{ m/s} - 0}{330 \text{ m/s} - 20.115 \text{ m/s}}\right)(445 \text{ Hz}) = 474 \text{ Hz} .$$

(b) For this part, the speeds of the source and receiver are each half of the above, $|v_s| = |v_r| = 10.058 \text{ m/s}$. Thus,

$$f' = \left(\frac{v - v_r}{v - v_s}\right)f_0 = \left(\frac{330 \text{ m/s} - (-10.058 \text{ m/s})}{330 \text{ m/s} - 10.058 \text{ m/s}}\right)(445 \text{ Hz}) = 473 \text{ Hz} .$$

What do you think? In both parts (a) and (b) the source and receiver move with respect to each other with the same relative velocity. Why are the answers slightly different? It's not round-off error.

Practice Quiz

8. A source of sound is moving away from a stationary observer at 15 m/s. If the frequency emitted by the source increases, the frequency heard by the receiver will

    (a) increase   **(b)** decrease   **(c)** stay the same   **(d)** none of the above

9. A source of sound is moving west at 15 m/s behind a receiver who is also moving west at 12 m/s. The frequency heard by the receiver, compared with the frequency emitted by the source, will be

    (a) lower   **(b)** higher   **(c)** the same   **(d)** none of the above

*14–9 Shock Waves

In the previous section we assumed $v_s < v$. When the source speed exceeds the speed of the wave in the medium, the sound emitted by the source becomes concentrated in a cone shaped region behind the source. This concentrated wave is called a *shock wave*. The size of the cone is determined by the ratio v/v_s called the **Mach number**.

Reference Tools and Resources

I. Key Terms and Phrases

transverse wave waves for which the oscillation is perpendicular to the direction of propagation

longitudinal wave waves for which the oscillation is along the direction of propagation

standing wave   a stationary wave resulting from the superposition of two waves traveling in opposite directions

wave number a quantity ($=2\pi/\lambda$) that measures the number of waves (in units of 2π rads) per unit length

wavelength the minimum repeat length of a wave

sound longitudinal vibrations of a medium

modes the allowed waves of a standing wave

fundamental frequency the frequency of the fundamental mode of a standing wave

harmonics the frequencies of the allowed standing waves

intensity the amount of energy per unit time per unit area

decibel scale a scale that measures the relative intensity of sound waves

Doppler effect the shift in frequency received due to relative motion between the source and the receiver

Mach number the ratio of the speed of the source of a sound to the speed of sound

II. Important Equations

Name/Topic	Equation	Explanation
The wave equation	$$\frac{\partial^2 z(x,t)}{\partial x^2} = \frac{1}{v^2}\frac{\partial^2 z(x,t)}{\partial t^2}$$	The equation that governs the behavior of a transverse wave on a string.
Wave properties	$$\lambda = 2\pi / k$$ $$v = \lambda f = \omega / k$$	The relationships between the wavelength, wave number, frequency, angular frequency, and speed of periodic waves.
Wave speeds	$$v = \sqrt{\frac{T}{\mu}} \; ; \; v = \sqrt{\frac{Y}{\rho}} \; ;$$ $$v = \sqrt{\frac{G}{\rho}} \; ; \; v = \sqrt{\frac{\gamma P_0}{\rho_0}}$$	The expressions for how the speeds of different waves depend on the medium.
Wave power	$$\langle P \rangle = \tfrac{1}{2}\mu\omega^2 A^2 v$$	The average power delivered by a mechanical wave.
Standing waves on a string	$$\lambda_n = \frac{2L}{n} \text{ and } f_n = \frac{n}{2L}\sqrt{\frac{T}{\mu}}$$	The wavelengths and frequencies of standing waves generated on strings with fixed ends.
Hearing sounds	$$\beta = 10\log_{10}\left(\frac{I}{I_0}\right)$$	The intensity level of a sound on the decibel scale.
Doppler effect	$$f' = \left(\frac{v - v_r}{v - v_s}\right)f_0$$	The relationship between the received (f') and emitted (f_0) frequencies when there is relative motion between the source and the receiver.

III. Know Your Units

Quantity	Dimension	SI Unit
Wavelength (λ)	$[L]$	m
Intensity (I)	$[MT^{-3}]$	W/m^2
Loudness (β)	dimensionless	dB

Practice Problems

1. A harmonic wave is given by $y(x,t) = 4\cos(3.53x - 6.41t)$. To the nearest hundredth of a meter, what is the wavelength?

2. What is its period in Problem 1, to the nearest thousandth of a second?

3. What is the velocity of the wave in Problem 1, to the nearest hundredth of a m/s?

4. What will be the value of y in problem 1 if $x = 0.9$ and $t = 1.47$, to two decimal places?

5. A point source of a wave emits 158 watts. To two decimal places, at a distance of 4.22 meters its intensity is ...

6. Two speakers emit sounds, the one with an intensity of 6.42×10^7 watts/m^2 and the other 7.57×10^7 watts/m^2. What is the difference in their intensity level, to the nearest tenth of a dB?

7. Taking right to be positive, if a speaker is moving at a **velocity** of 24 m/s while emitting a sound of frequency 1438 Hz, what frequency will you hear?

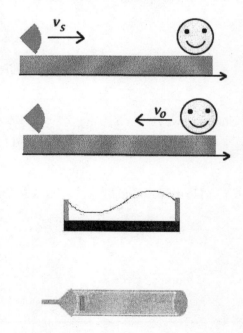

8. If the situation is reversed in Problem 7 and you are the one that is moving, what would the answer be? (Remember right is positive.)

9. A vibrating stretched string has length 29 cm, mass 18 grams and is under a tension of 44 newtons. What is the frequency of its 3rd harmonic, to the nearest Hz?

10. If the source of the wave in Problem 9 was an open organ pipe of the same length as the wire, what would the frequency of the 5th harmonic be?

Selected Solutions to End of Chapter Problems

17. We find the speed of the wave from

$$v = (Y/\rho)^{1/2} = [(2.10 \times 10^{11} \text{ N/m}^2)/(7.9 \times 10^3 \text{ kg/m}^3)]^{1/2} = 5.16 \times 10^3 \text{ m/s}.$$

The time for the message to travel along the cable is

$$t = L/v = (5.0 \times 10^3 \text{ m})/(5.16 \times 10^3 \text{ m/s}) = 0.97 \text{ s}.$$

$\boxed{\text{Yes}}$, the message does arrive in time!

29. For the intensity, we have $I = P/\text{area} = \frac{1}{2}\rho\omega^2 A^2 v = \frac{1}{2}4\pi^2\rho f^2 A^2 v$;

Solving for the area gives,

$$A = \left(\frac{2I}{4\pi^2\rho v f^2}\right)^{1/2} = \left[\frac{2\left(5.0 \times 10^{-7}\,\frac{\text{W}}{\text{m}^2}\right)}{4\pi^2\left(1.30\,\frac{\text{kg}}{\text{m}^3}\right)\left(330\,\frac{\text{m}}{\text{s}}\right)(8600 \text{ Hz})}\right]^{1/2} = \boxed{8.9 \times 10^{-10}\,\text{m}}$$

This length is comparable to the size of molecules.

41. Let string 1 be the thinner string and string 2 be the thicker string. We are told that string 2 is 3 times thicker, which means that it has 3 times the diameter: $D_2 = 3D_1$. The volume of each string is given by the cross sectional area times the length: $V = (\pi D^2/4)L$. The linear density of a string is its mass per unit length, where the mass is given by the product of volume density and volume. So, using the fact that the two strings are both made of steel and therefore have the same volume density, we can write

$$\frac{\mu_2}{\mu_1} = \frac{m_2/L}{m_1/L} = \frac{\rho V_2}{\rho V_1} = \frac{L\left(\pi D_2^2/4\right)}{L\left(\pi D_1^2/4\right)} = \left(\frac{D_2}{D_1}\right)^2 = (3)^2 = 9$$

Because the strings have the same length and are each in the fundamental mode, the wavelengths are the same. So, for the ratio of frequencies, we have

$$\frac{f_2}{f_1} = \frac{\sqrt{T_2/\mu_2}}{\sqrt{T_1/\mu_1}} = \sqrt{\frac{T_2}{T_1}\cdot\frac{\mu_1}{\mu_2}} \quad\Rightarrow\quad \frac{T_2}{T_1} = \frac{\mu_2}{\mu_1}\left(\frac{f_2}{f_1}\right)^2 = 9\left(\frac{400 \text{ Hz}}{1200 \text{ Hz}}\right)^2 = \boxed{1}.$$

49. From $\beta = 10\log_{10}(I/I_0)$, we get

$\beta_1 = 10\log_{10}[(8.3 \times 10^{-4} \text{ W/m}^2)/(10^{-12} \text{ W/m}^2)] = \boxed{89 \text{ dB}}$;

$\beta_2 = 10\log_{10}[(3.5 \times 10^{-6} \text{ W/m}^2)/(10^{-12} \text{ W/m}^2)] = \boxed{65 \text{ dB}}$;

$\beta_3 = 10\log_{10}[(7.2 \times 10^{-9} \text{ W/m}^2)/(10^{-12} \text{ W/m}^2)] = \boxed{39 \text{ dB}}$.

59. The speed of the train is $(140 \text{ km/h}) = (140{,}000 \text{ m/h}) \times (1\text{h}/3600 \text{ s}) = 38.9 \text{ m/s}.$

(a) Because the wavelength in front of a moving source decreases, the wavelength traveling toward the observer is

$$\lambda = (v - v_T)/f_s = (330 \text{ m/s} - 39 \text{ m/s})/(333 \text{ Hz}) = \boxed{0.874 \text{ m}}.$$

This wavelength approaches the observer at a speed of v. He hears a frequency

$$f_a = v/\lambda = (330 \text{ m/s})/(0.874 \text{ m}) = \boxed{378 \text{ Hz}}.$$

(b) The wavelength traveling toward the observer is

$$\lambda_b = v/f_T = (330 \text{ m/s})/(333 \text{ Hz}) = 0.991 \text{ m}.$$

This wavelength approaches the observer at a relative speed of $v + v_O$. He hears a frequency

$$f_b = (v + v_O)/\lambda_b = (330 \text{ m/s} + 39 \text{ m/s}) / (0.991 \text{ m}) = \boxed{372 \text{ Hz}}.$$

79. The enhancement of the sound intensity occurs when a standing wave is produced in the air column. Because the water vibration is much smaller than the air vibration, the top of the water column can be considered a fixed point for the standing wave in the air column. Adjacent fixed points are separated by $\lambda/2$, so the wavelength is

$$\lambda = 2(1.0 \text{ m} - 0.60 \text{ m}) = 0.80 \text{ m}.$$

The speed of sound in the air is

$$v = f\lambda = (440 \text{ Hz})(0.80 \text{ m}) = \boxed{3.5 \times 10^2 \text{ m/s}}.$$

Answers to Practice Quiz

1. (b) **2.** (d) **3.** (a) **4.** (c) **5.** (c) **6.** (d) **7.** (e) **8.** (a) **9.** (b)

Answers to Practice Problems

1.	1.78 m	**2.**	0.980 s
3.	1.82 m/s	**4.**	4.00 m
5.	0.71 W/m^2	**6.**	0.7 dB
7.	1551 Hz	**8.**	1333 Hz
9.	138 Hz	**10.**	2845 Hz

CHAPTER 15

SUPERPOSITION AND INTERFERENCE OF WAVES

Chapter Objectives

After studying this chapter, you should

1. understand how the principle of superposition applies to wave behavior.
2. understand the difference between constructive and destructive interference.
3. be able to determine the beat wavelength and the beat frequency.
4. understand the spatial interference from two sources.
5. understand the reflection and transmission of wave pulses.
6. understand the basic ideas behind the Fourier decomposition of waves.

Chapter Review

In this chapter, we continue our study of waves. Our focus here is on phenomena that can be understood by the principle of superposition. This principle tells us how waves combine; and through this principle, we have come to understand a lot of important aspects of wave behavior.

15–1—15–2 Superposition and Standing Waves

The **superposition principle** is the fact that if two or more waves exist in the same place at the same time, the result is a wave that is the sum of the displacements of the individual waves

$$z(x,t) = z_1(x,t) + z_2(x,t) + \cdots$$

Mathematically, this principle is a consequence of the fact that the wave equation is linear. The physical addition of waves is called wave *interference*.

Characteristic of wave interference are regions where two special cases can be identified. One of these special cases occurs when the individual waves add in such a way that their maxima and/or their minima are at the same place at the same time. The result of this occurrence is that the amplitude of the resultant wave equals the sum of the amplitudes of the individual waves. This effect is called **constructive interference**. The other special case, **destructive interference**, occurs when the maximum of one wave meets the minimum of another. In this latter case, the amplitude of the resultant wave equals the difference of the amplitudes of the individual waves.

Wave interference can sometimes produce a consistent interference pattern, but only if the waves have fixed relations between their frequencies and phases. Such waves are called *coherent*. Where there are no such fixed relations, the waves are said to be *incoherent* and produce no consistent interference pattern.

Standing waves result from the superposition of two identical waves traveling in opposite directions. If $z_r = A\sin(kx - \omega t)$ is a wave traveling to the right, and $z_l = A\sin(kx + \omega t)$ is an identical wave traveling to the left, then their superposition produces the wave

$$z_1 = z_r + z_l = 2A\sin(kx)\cos(\omega t).$$

The familiar way to construct this situation is for one of the waves to be the reflection of the other.

15–3 Beats

The superposition of waves of different frequencies gives rise to the phenomenon called **beats**. These beats appear as a slower regular fluctuation in the amplitude of the wave that results from the superposition. This more slowly varying lower frequency wave corresponds to the beats. The frequency and wavelength of this wave are called the *beat frequency* and *beat wavelength*, respectively. Two waves of frequencies f_1 and f_2 (and wavelengths λ_1 and λ_2) will produce a beat frequency and a beat wavelength given by

$$f_{beat} = \frac{1}{2}(f_1 - f_2)$$

$$\frac{1}{\lambda_{beat}} = \frac{1}{2}\left(\frac{1}{\lambda_1} - \frac{1}{\lambda_2}\right),$$

assuming that $f_1 > f_2$. It is important to note that with sound, the perceived beat frequency is twice the above result because the ear will hear an increased intensity of the sound for both the maximum and the minimum of the amplitude variation. Beats are often heard when two guitar strings are played at the same time and are often used to tune musical instruments to the desired frequency.

Example 15–1 A string plays a fundamental frequency of 448 Hz. If matched against a 440 Hz tuning fork, what beat frequency will be perceived by the listener?

Setting it up:
The information provided in the problem is the following:
Given: f_1 = 448 Hz, f_2 = 440 Hz;   **Find:** $f_{perceived}$

Strategy:
Here we must consider that the ear perceives beats of sound at twice the frequency of the wave envelope.

Working it out:
We know that

$$f_{perceived} = 2f_{beat} = f_1 - f_2 = 448\ \text{Hz} - 440\ \text{Hz} = 8\ \text{Hz}.$$

What do you think?   What is the angular frequency of the perceived beats?

Practice Quiz

1. If sound travels at 330 m/s, what is the beat wavelength for the situation in Example 15–1?
    (a) 0.737 m   **(b)** 0.750 m   **(c)** 82.5 m   **(d)** 41.2 m/s   **(e)** 165 m

2. What angular frequency is associated with the actual beat wave for the situation in Example 15–1?
    (a) 4 rad/s   **(b)** 25 rad/s   **(c)** 8 rad/s   **(d)** 50 rad/s   **(e)** 0.6 rad/s

15–4 Spatial Interference Phenomena

As we've previously discussed, when more than one wave passes through the same medium simultaneously the superposition of these waves amounts to the addition of the individual displacements to form a single wave. If the individual waves are coherent their superposition produces a spatial interference pattern. Certain positions in this pattern will correspond to constructive interference and others to destructive interference. Whether the interference is constructive or destructive at a certain location depends on the difference in the length of the path, ΔL, taken by the individual waves. The conditions are

$$\text{constructive interference: } \Delta L = n\lambda, \ n = 0, \pm 1, \pm 2, \ldots$$

$$\text{destructive interference: } \Delta L = (n + \tfrac{1}{2})\lambda, \ n = 0, \pm 1, \pm 2, \ldots$$

This means that constructive interference occurs when the path difference is an integer multiple of the wavelength of the waves and destructive interference occurs when it is shifted by one-half wavelength from the condition for constructive interference.

For two coherent sources of wavelength λ, separated by a distance d, the above conditions reduce to

$$\text{constructive interference: } \sin\theta = n\frac{\lambda}{d}, \ n = 0, \pm 1, \pm 2, \ldots$$

$$\text{destructive interference: } \sin\theta = (n + \tfrac{1}{2})\frac{\lambda}{d}, \ n = 0, \pm 1, \pm 2, \ldots$$

where θ is the approximately equal angle from either source to the point a distance $R \gg d$ away from the sources (see Fig. 15–10 (b) on p. 444 in the textbook) where the condition is being checked.

Example 15–2 Each of two point sources a distance d = 3.0 m apart emits sound waves of equal wavelength with opposite phase. If a detector placed 5.2 m from the line joining the sources and one-quarter of the way from one to the other, detects constructive interference between the two waves, what is the maximum possible wavelength of the sound wave?

Setting it up: A sketch is helpful. The diagram shows the two sources (top and bottom). The detector is indicated as being a distance R from the line joining the sources and 1/4 of the way between them from the top source. The information given in the problem is the following:
Given: d = 3.0 m, R = 5.2 m; **Find:** λ_{max}

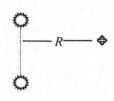

Strategy:

We need to determine the path difference between waves from each source in terms of the distance R. We must then correctly identify how this path difference relates to the wavelength to produce constructive interference.

Working it out:

The path length of the top source is the hypotenuse of the right triangle with legs R and $d/4$

$$L_{\text{Top}} = \sqrt{R^2 + (d/4)^2} = \sqrt{(5.2 \text{ m})^2 + \left(\frac{3.0 \text{ m}}{4}\right)^2} = 5.254 \text{ m}.$$

The path length of the bottom source is the hypotenuse of the right triangle with legs R and $\frac{3}{4}d$

$$L_{\text{Bott}} = \sqrt{R^2 + \left(\frac{3d}{4}\right)^2} = \sqrt{(5.2 \text{ m})^2 + \left(\frac{9.0 \text{ m}}{4}\right)^2} = 5.666 \text{ m}.$$

The path difference between the two sources is

$$\Delta L = L_{\text{Bott}} - L_{\text{Top}} = 5.667 \text{ m} - 5.254 \text{ m} = 0.413 \text{ m}$$

For constructive interference, the path difference must be an integral multiple of $\frac{1}{2}\lambda$ because the waves start with opposite phases. Thus, the maximum possible wavelength is

$$\Delta L = \frac{\lambda_{\text{max}}}{2} \quad \Rightarrow \quad \lambda_{\text{max}} = 2(\Delta L) = 2(0.413 \text{ m}) = 0.83 \text{ m}.$$

What do you think?   The answer gives only one possible wavelength. What are some others?

Practice Quiz

3. Suppose two sources of identical waves emit them in phase. When these waves superimpose in the surrounding space, which of the following is a condition on the path difference, ΔL, between the waves that will produce destructive interference?
  (a) $\Delta L = \lambda$ **(b)** $\Delta L = \lambda/2$ **(c)** $\Delta L = \lambda/3$ **(d)** $\Delta L = \lambda/4$ **(e)** none of the above

15–5 Pulses

Many of the important properties of waves can be understood by studying the behavior of **pulses**. A pulse is an isolated disturbance that propagates through a medium. Some of the basic facts concerning the behavior of pulses follow:

- Collisions: When two pulses collide they obey the superposition principle. The displacements add as they pass through each other. Each pulse continues to move even when there is complete destructive interference momentarily.
- Reflection: When a pulse reaches the boundary of the medium in which it travels, there is reflection off the boundary. For a pulse on a string, if the boundary is a fixed end the reflected pulse has an inverted displacement relative to the incident pulse. If the boundary of the string is a free end, the displacement of the reflected pulse is in the same direction as the incident pulse.
- Transmission: When a pulse reaches the boundary between its incident medium and another medium there is also transmission into the second medium. For a pulse on a string, the displacement of the transmitted pulse is the same as that of the incident pulse. However, the displacement of the reflected pulse is inverted if the second string is more dense than the incident string and remains the same if the second string is less dense. By conservation of energy, the amplitude of the incident pulse must be greater than that of the reflected pulse and that of the transmitted pulse.

- Velocity: Because the speed of a pulse on a string depends on the density of the string, the speeds of the incident and reflected pulses will be different from that of the transmitted pulse. The transmitted pulse will move faster if the second medium is of lower density and slower if its density is higher.

Example 15–3 The two pulses shown below move in opposite directions as indicated. Each pulse has a speed of 1 cm/s. What disturbance is in the region one second later?

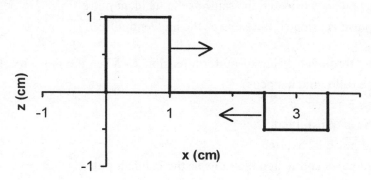

Setting it up:

The given figure sets up the problem nicely. Let's call the left pulse P_1 and the right pulse P_2.
Given: $v = 1$ cm/s, $t = 1$ s;   **Find:** wave at time t

Strategy:

Wherever there is overlap of the two pulses, the displacements will add.

Working it out:

After 1 second, P_1 spans the region from $x = 1$ cm to $x = 2$ cm; and P_2 spans the region from $x = 1.5$ cm to $x = 2.5$ cm. They overlap in the region from $x = 1.5$ cm to $x = 2$ cm. Within this overlapping region, the amplitude of the resultant pulse will be 1 cm – 0.5 cm = 0.5 cm. Outside of the overlapping region, the individual pulses will look the same as they do above. Thus, we have the following result.

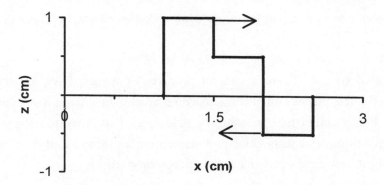

What do you think? How will the region look after another second passes?

Practice Quiz

4. A pulse is traveling through a compound string of uniform tension T. When the pulse reaches a boundary traveling from a lower density part of the string, which of the following happens?

    (a) The entire pulse is reflected.

    (b) The entire pulse is transmitted.

    (c) The transmitted pulse is inverted.

    (d) The transmitted pulse is oriented the same as the incident pulse.

    (e) The reflected pulse is oriented the same as the incident pulse.

5. A pulse is traveling through a string of uniform tension T. When the pulse reaches a fixed end of the string, which of the following happens?

    (a) The entire pulse is reflected.

    (b) The entire pulse is transmitted.

    (c) The transmitted pulse is inverted.

    (d) The transmitted pulse is oriented the same as the incident pulse.

    (e) The reflected pulse is oriented the same as the incident pulse.

15–6—15–7 Fourier Decomposition of Waves and the Uncertainty Principle

One of the reasons the study of sinusoidal waves is so important is because we know that any periodic wave or pulse can be approximated by a superposition of sinusoidal waves. The individual waves that make up this superposition can be determined and is referred to as a **Fourier decomposition** of the approximated wave. Both the time dependence and the spatial properties of the wave can be approximated in this way.

Through Fourier decomposition, we know that a pulse that is either limited in time and/or space contains a limited number of frequencies and/or wave numbers. These facts are represented by reciprocal relations. For a pulse that lasts for a time Δt, the range of frequencies in its Fourier decomposition, Δf, is related to it by

$$(\Delta t)(\Delta f) \geq 1.$$

A pulse of spatial width Δx consists of a Fourier decomposition of waves whose wave number range, Δk, is related to it by

$$(\Delta x)(\Delta k) \geq 2\pi.$$

These reciprocal relations, which are a general property of waves, have an important application in quantum mechanics – the physics of the microscopic world. In quantum mechanics, these reciprocal relations lead to what is called the **uncertainty principle**. This principle states that the precision of certain conjugate quantities, such as position and momentum, cannot be known to arbitrary accuracy. This fact leads to a lot of interesting phenomena in the microscopic world.

Reference Tools and Resources

I. Key Terms and Phrases

superposition principle the resultant of two or more waves moving through the same medium at the same time is the sum of the displacements of the individual waves

constructive interference when waves superimpose in phase resulting in a wave of larger amplitude

destructive interference when waves superimpose out of phase resulting in a wave of smaller amplitude

beats variation in the intensity of a wave resulting from the superposition of waves of different frequency

pulse an isolated disturbance propagating through a medium

Fourier decomposition the process of approximating an arbitrary wave as a superposition of simple sinusoidal waves

uncertainty principle reciprocal relations between the uncertainties in physical quantities in the context of quantum physics

II. Important Equations

Name/Topic	Equation	Explanation
Standing waves	$z(x,t) = 2A\sin(kx)\cos(\omega t)$	The standing wave solution from the interference of two identical waves traveling in opposite directions.
Beats	$\dfrac{1}{\lambda_{beat}} = \dfrac{1}{2}\left(\dfrac{1}{\lambda_1} - \dfrac{1}{\lambda_2}\right)$ $f_{beat} = \tfrac{1}{2}(f_1 - f_2)$ $f_{perceived} = 2f_{beat}$	The wavelength and frequency of the beats that result from two different waves moving through the same medium.
Spatial interference	$\Delta L = \begin{cases} n\lambda \\ \left(n + \tfrac{1}{2}\right)\lambda \end{cases} \quad n = 0, \pm 1, \pm 2, \ldots$	The conditions for constructive (upper) and destructive (lower) interference at a point where identical waves from separate sources meet.
Fourier decomposition	$f(t) = \displaystyle\sum_{n=1}^{\infty} A_n \sin(n\omega t + \phi_n)$ $g(x) = \displaystyle\sum_{n=1}^{\infty} A'_n \sin(nkx + \theta_n)$	The temporal and spatial sums that are used to approximate an arbitrary wave from sinusoidal waves.
Reciprocal relations	$(\Delta t)(\Delta f) \geq 1$ $(\Delta x)(\Delta k) \geq 2\pi$	The reciprocal relations for the Fourier decomposition of a wave pulse.

Practice Problems

1. Two sinusoidal waves traveling along a string with equal amplitudes and with periods of 18 and 20 seconds, respectively, are observed to have a maximum displacement at $x = 0$ and $t = 0$. To the nearest second, after how many seconds will the first beat minimum occur?

2. Two sounds have a frequency of 1002 and 1012 Hz, respectively. To one decimal place what is the beat frequency?

3. In the previous problem, how many beats are heard per second?

4. The speed of the propagation of light in different media depends on the properties of the media. The formula is given by $n = c/v$ where $c = 3 \times 10^8$ m/s and n is the index of refraction ($n > 1$ for all media). Consider light of frequency 0.83×10^{15} Hz entering from empty space, with $n = 1$ into a media, with $n = 1.33$. To the nearest tenth of a nanometer, what is the wavelength of the light in empty space?

5. In the previous problem, what is the wavelength of the light in glass?

6. Two strings are tied together. The first has a mass density of 2.9 gm/cm and the second 6 gm/cm. A wave traveling on the first string is partially reflected with an amplitude of 2.15 cm and partially transmitted with an amplitude of 2.14 cm. To two decimal places, what was the amplitude of the wave?

7. Two loudspeakers are separated by 0.5 m. These speakers form sound waves of the same amplitude at a frequency of 6844 Hz. The amplifier emits the two waves 180° out of phase. A set of chairs is arranged in a semicircle 35 m from the midpoint of the two speakers. To two decimal places, at what distance to the right of the central chair is there a first maximum in sound intensity? (You can use the small angle approximation.)

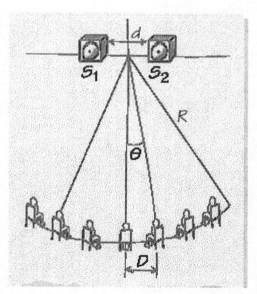

8. In the previous problem, at what distance to the right of the central chair is there a first minimum in sound intensity? (You can use the small angle approximation.)

9. A synthesizer, which is an electronic device for generating and combining harmonic waves, generates an electromagnetic pulse of width 0.72×10^{-6} s. To the nearest Hz, what is the range of frequencies that the synthesizer must utilize to generate the pulse?

10. Two strings of the same material on a pair of guitars are in tune with a frequency of 408 Hz. The string of one of them is tightened slightly, and you hear beats at a frequency of 7 Hz. What is the frequency of the tightened string?

Selected Solutions to End of Chapter Problems

5. We can write the combined wave, with $A = 1$ cm, as

$$\psi(x, t) = 3A \sin(kx - \omega t) + 2A \sin(kx - \omega t - \pi/2) = 3A \sin(kx - \omega t) + 2A \cos(kx - \omega t).$$

We want to express this as a single wave

$$\psi(x, t) = B \sin(kx - \omega t + \alpha)$$

For the sake of comparison we note that this single wave can be written as

$$\psi(x, t) = B \sin(kx - \omega t + \alpha) = B \sin(kx - \omega t) \cos\alpha + B \cos(kx - \omega t) \sin\alpha.$$

Comparing coefficients, we have

$$3A = B \cos \alpha \qquad \text{and} \qquad 2A = B \sin \alpha,$$

which gives

$$\tan\alpha = \tfrac{2}{3} \quad \Rightarrow \quad \alpha = 33.7°$$

Therefore,

$$B = 3A/\cos\alpha = 3(1 \text{ cm})/\cos(33.7°) = \boxed{3.6 \text{ cm}}.$$

11. (a) The frequencies and speeds are the same:

$$f = \omega/2\pi = (22\pi \text{ s}^{-1})/2\pi = \boxed{11 \text{ Hz}};$$

$$v = \omega/k = (22\pi \text{ s}^{-1})/(3\pi \text{ m}^{-1}) = \boxed{7.3 \text{ m/s}}.$$

(b) A standing wave of the form given is produced by identical waves traveling in opposite directions of the general form $z = A \sin(kx \pm \omega t)$, where A is the amplitude of the individual waves. This amplitude works out to be half the amplitude of the standing wave. Therefore, the waveform of the two waves in question are

$$z_\ell = \boxed{3 \sin(kx + \omega t)} \qquad \text{and} \qquad z_r = \boxed{3 \sin(kx - \omega t)}$$

15. For a standing wave, the distance between nodes must be a multiple of $\lambda/2$. The largest possible wavelength is $\lambda/2 = 6$ m $- 3$ m, which gives $\lambda = \boxed{6 \text{ m}}$. There is a node at $x = 0$. Because the first wave has a maximum here at $t = 0$, the second wave must have a minimum here at $t = 0$.

21. Because the 4 beats/s are heard, this must be the pulse frequency. With the small fractional change, we approximate the changes by differentials:

$$f = \frac{v}{\lambda} = \frac{1}{\lambda}\left(\frac{T}{\mu}\right)^{1/2} \quad \Rightarrow \quad df = \frac{1}{2\lambda}\left(\frac{1}{T\mu}\right)^{1/2} dT$$

Taking the ratio df/f, we get

$$\frac{df}{f} = \frac{1}{2}\frac{dT}{T} \quad \Rightarrow \quad \frac{dT}{T} = 2\frac{df}{f}$$

Therefore,

$$\frac{dT}{T} = 2\left(\frac{-4\text{ Hz}}{512\text{ Hz}}\right) = -0.016 = \boxed{-1.6\%}$$

25. The angle of the next maximum is given by
$$\tan\theta = D/L = 8.0\text{ cm}/40.0\text{ cm} = 0.20,$$
which gives $\theta = 11.3°$. From the analysis of Figure 15–11, we know that the path-length difference from two sources separated by d to the point an angle θ from the centerline is
$$\Delta L = d\sin\theta.$$
For the next maximum, this path-length difference must be λ:
$$\lambda = d\sin\theta = (3.0\text{ cm})\sin(11.3°) = \boxed{0.59\text{ cm}}.$$

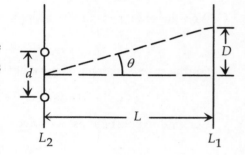

53. The angular frequency of the turntable is $\omega = (120\text{ rev/min})(2\pi\text{ rad/rev})(1\text{ min}/60\text{ s}) = 4\pi\text{ rad/s}$. The tangential speed of the sources is

$$v_t = R\omega = (0.50\text{ m})(4\pi\text{ rad/s}) = 2\pi\text{ m/s}.$$

If we take $t = 0$ when a source has a tangential velocity directly toward the detector, the wavelength from the approaching source is

$$\lambda_1 = [v - v_t\cos(\omega t)]/f_s$$

and the frequency detected will be

$$f_1 = v/\lambda = vf_s/[v - v_t\cos(\omega t)].$$

The frequency from the second source can be obtained by adding π rad to the angle:

$$f_2 = vf_s/[v - v_t\cos(\omega t + \pi)] = vf_s/[v + v_t\cos(\omega t)].$$

The pulse frequency is

$$f_{\text{pulse}} = f_1 - f_2 = vf_s/[v - v_t\cos(\omega t)] - vf_s/[v + v_t\cos(\omega t)]$$
$$= f_s(\, 1/[1 - (v_t/v)\cos(\omega t)] - 1/[1 + (v_t/v)\cos(\omega t)]\,).$$

Because $v_t \ll v$, we can use the approximation $1/(1 - x) \approx 1 + x$:

$$f_{\text{pulse}} \approx f_s[1 + (v_t/v)\cos(\omega t) - 1 + (v_t/v)\cos(\omega t)]$$
$$= 2f_s(v_t/v)\cos(\omega t) = 2(1800\text{ Hz})[(2\pi\text{ m/s})/(330\text{ m/s})]\cos(4\pi t)$$
$$= \boxed{69\cos(4\pi t)\text{ Hz}}.$$

Answers to Practice Quiz

1. (c) 2. (b) 3. (b) 4. (d) 5. (a)

Answers to Practice Problems

1. 90 s	2. 5.0 Hz
3. 10	4. 361.4 nm
5. 271.8 nm	6. 3.35 cm
7. 1.69 m	8. 3.38 m
9. 1,388,889 Hz	10. 415 Hz

CHAPTER 16

PROPERTIES OF FLUIDS

Chapter Objectives

After studying this chapter, you should

1. be able to identify the states of matter.
2. be able to calculate average density and pressure.
3. be able to determine the fluid pressure at various depths.
4. be able to use Archimedes' principle to determine the buoyant force on an object.
5. be able to use the continuity equation and Bernoulli's equation to describe the behavior of ideal fluids.
6. be able to account for viscosity and turbulence in real fluids.

Chapter Review

In this chapter, we study **fluids**. Fluids are characterized by their ability to flow; both liquids and gases are considered to be fluids. An understanding of fluid behavior is essential to life, and applications of this understanding are essential to many of the conveniences of modern living.

16–1—16–3 States of Matter, Density, & Pressure

There are three primary states of matter: solids, liquids, and gases (sometimes a fourth state called *plasma* is added). In the gaseous state, molecules are typically far enough apart that the forces between them are negligible except during brief collisions between the molecules. In solids and liquids, the intermolecular forces are strong enough to hold the molecules close to each other. However, with solids these forces are resistance to both compressive and shear deformations, whereas with liquids there is very little resistance to shearing. This means that liquids (and gases) will easily flow, but solids will not. Hence, liquids and gases are called fluids.

One of the most convenient properties used to describe a fluid is its *density*, ρ. The density of a substance is a measure of how compact the substance is, that is, how much mass is packed into a volume of the substance. On average, the density of a substance is the amount of mass M divided by the volume V taken up by that mass

$$\rho = \frac{M}{V}$$

The density of a substance is also commonly measured relative to the density of liquid water, which is $\rho_w = 1.00 \times 10^3$ kg/m^3. The ratio of the density of a substance to the density of water, ρ/ρ_w, is called its **specific gravity**.

Another important quantity in the study of fluids is *pressure*, p. On average, the pressure that is applied to an object is the amount of force F (normal to the surface) divided by the area A over which the force spreads

$$p = \frac{F}{A}$$

As the preceding expressions indicate, the unit of pressure is that of force divided by area; in SI units this is called the pascal (Pa): 1 Pa = 1 N/m². Other common units of pressure are pounds per square inch, atmospheres (conversion factors are given in the text). Two important properties of the pressure in a static fluid are that it is equally applied in all directions (at a given depth) and that the applied forces are perpendicular to any surface in the fluid.

In the case of static fluids, every part of the fluid and every object within the fluid is in static equilibrium. One of the basic properties of fluids is known as **Pascal's principle**:

A change in pressure applied to an enclosed fluid is transmitted undiminished throughout the fluid.

Pascal's principle is important in determining the dependence of pressure on the depth within a fluid. Without any external pressure applied to the outer surface of a fluid, the pressure measured at a depth y beneath the surface arises from the weight of the fluid above the given level. The amount of this increase in pressure is given by $\rho g y$. If there is external pressure on the fluid, such as from the atmosphere and/or any other source, this pressure is transmitted undiminished to every point in the fluid and it must be added to the pressure due to the weight of the fluid. Thus, for the dependence of pressure on depth we have

$$p = p_0 + \rho g y$$

where p is the pressure at a given level and p_0 is the pressure at a height y above that level.

Example 16–1 A uniform cylindrical container has a radius of 7.8 cm and a height of 13.2 cm. If this container is completely filled with water, what pressure does the water apply to the bottom?

Setting it up:
The information provided in the problem is the following:
Given: $r = 7.8$ cm, $h = 13.2$ cm; **Find:** p_{bott}

Strategy:
The pressure applied by the water is the force that the water applies (equal to its weight) divided by the bottom area of the cylinder.

Working it out:
The weight of the water is given by mg; to determine the mass of water, we will need the volume of water. The volume of water equals the volume of the cylinder, $V = Ah$. So, the mass of water is given by $m = \rho_w V = \rho_w Ah$. The pressure on the bottom then, is given by

$$p = \frac{mg}{A} = \frac{(\rho_w Ah)g}{A} = \rho_w hg \ .$$

This gives

$$p = \left(1000 \text{ kg/m}^3\right)\left(0.132 \text{ m}\right)\left(9.8 \text{ m/s}^2\right) = 1.3 \times 10^3 \text{ Pa} \ .$$

What do you think?   What would the pressure be if you account for atmospheric pressure?

Example 16–2 Human blood has a density of approximately 1.05×10^3 kg/m^3. Use this information to estimate what the difference in blood pressure would be between the brain and the feet of a person who is approximately 6 feet tall if the blood were a static fluid.

Setting it up:

The information provided in the problem is the following:

Given: $y = 6$ ft, $\rho = 1.05 \times 10^3$ kg/m^3;   **Find:** Δp

Strategy:

We attempt this approximation, to two significant digits, using the result for the dependence of pressure on depth.

Working it out:

The pressure-depth equation gives,

$$p = p_0 + \rho g y \quad \Rightarrow \quad p - p_0 = \Delta p = \rho g y .$$

Therefore,

$$\Delta p = \left(1.05 \times 10^3 \text{ kg/m}^3\right)\left(9.8 \text{ m/s}^2\right)\left(6.0 \text{ ft}\right)\left(\frac{1 \text{ m}}{3.28 \text{ ft}}\right) = 1.9 \times 10^4 \text{ Pa} .$$

What do you think?   The blood is not a static fluid in the body, so this is only a ballpark estimate. See if you can look up a more realistic approximation for this difference.

Practice Quiz

1. If it takes twice as much volume of fluid 1 to weigh the same as fluid 2, how do their densities compare?
   (a) Fluid 1 is twice as dense as fluid 2.
   (b) Fluid 1 is half as dense as fluid 2.
   (c) Fluid 1 and fluid 2 have equal densities.
   (d) Fluid 1 is four times less dense than fluid 2.
   (e) None of the above.

2. If a uniform cylinder of height 0.850 m and radius 0.250 m is completely filled with water, what is the *total* pressure on the bottom of the cylinder?
   (a) 8.34 kPa   (b) 133 kPa   (c) 2.08 kPa   (d) 109 kPa   (e) 46.2 kPa

3. What is the pressure, from gravity, 3.4 m below the surface of a container filled with a fluid of density 550 kg/m^3?
   (a) 33 kPa   (b) 1900 Pa   (c) 18 kPa   (d) 100 kPa   (e) 550 Pa

4. The pressure at a particular location in a fluid of density 870 kg/m³ is 120 kPa. What is the pressure in the fluid 5.9 m above this location?

   (a) 70 kPa   **(b)** 50 kPa   **(c)** 62 kPa   **(d)** 58 kPa   **(e)** 20 kPa

16–4 Buoyancy and Archimedes' Principle

When an object is submerged in a fluid, the volume taken up by the object displaces an equal volume of the fluid. The pressure applied by the fluid onto the object results in an upward force on the object; this phenomenon is known as **buoyancy**. Our understanding of buoyancy is governed by **Archimedes' principle**:

> *An object immersed in a fluid experiences an upward force equal to the weight of the fluid displaced by the object.*

The weight of the fluid displaced by the object equals the mass of this fluid times the acceleration due to gravity, $m_{flu}g$. When dealing with buoyancy, it is usually more convenient to express the mass in terms of the density, $m_{flu} = \rho_{flu}V$, where V is the volume of fluid displaced. Therefore, for an object submerged in a fluid, the buoyant force on it is

$$F_b = \rho_{flu}gV_{sub},$$

where V_{sub} is the volume of the object that is submerged in the fluid.

   The net vertical force on a completely submerged object, given by the difference between its weight and the buoyant force, is

$$F_{net} = F_g - F_b = \rho_{obj}gV - \rho_{flu}gV = \left(\rho_{obj} - \rho_{flu}\right)gV.$$

So, this net force will either be positive (downward in this case) or negative (upward) depending on whether the object is more or less dense than the fluid. This result represents the well known fact that objects that are more dense than water will sink while objects that are less dense than water will float.

Example 16–3 Many magic tricks are based on physical principles. In order to fool her audience, a magician uses an object that sinks in the freshwater made available to the audience, but floats in the seawater that she uses on stage. If the density of seawater is 1025 kg/m³, what maximum percentage of the object's volume will float above the seawater?

Setting it up:
The information provided in the problem is the following:
Given: $\rho_{sea} = 1.025 \times 10^3$ kg/m³;   **Find:** percentage of volume above seawater

Strategy:
As a result of Archimedes' principle, we know that more of an object will be submerged if its density approaches that of the water, so we get the maximum above-surface float for the smallest possible object density. Since it must completely submerge in fresh water, the smallest object density is 1.00×10^3 kg/m³. We are interested in the case where the object floats in equilibrium, so the net force on it is zero.

Working it out:

The net force can be written out for this case as

$$F_{net} = F_g - F_b = \rho_{obj} g V_{obj} - \rho_{sea} g V_{sub} = 0 .$$

Therefore,

$$\rho_{obj} g V_{obj} = \rho_{sea} g V_{sub} \quad \Rightarrow \quad \frac{V_{sub}}{V_{obj}} = \frac{\rho_{obj}}{\rho_{sea}} .$$

Now, the volume submerged is related to the volume above by $V_{sub} + V_{above} = V_{obj}$. So, we can write

$$\frac{V_{obj} - V_{above}}{V_{obj}} = 1 - \frac{V_{above}}{V_{obj}} = \frac{\rho_{obj}}{\rho_{sea}} \quad \therefore \quad \frac{V_{above}}{V_{obj}} = 1 - \frac{\rho_{obj}}{\rho_{sea}} .$$

This gives

$$\frac{V_{above}}{V_{obj}} = 1 - \frac{1000 \text{ kg/m}^3}{1025 \text{ kg/m}^3} = 0.0244$$

which tells us that 2.44% of the volume floats above the seawater.

What do you think?   What could the magician do to the seawater to make a larger percentage of the object float?

Practice Quiz

5.   An object of density 750 kg/m^3 is half submerged in a fluid. What is the density of this fluid?

  (a) 1500 kg/m^3 **(b)** 188 kg/m^3 **(c)** 375 kg/m^3 **(d)** 2250 kg/m^3 **(e)** 750 kg/m^3

6.   If the volume of the object in question 5 is 0.33 m^3, what is the buoyant force on this object?

  (a) 9810 N   **(b)** 1210 N   **(c)** 3240 N   **(d)** 1000 N   **(e)** 2430 N

16–5—16–8 Fluids in Motion, Continuity, & Bernoulli's Equation

In this section, we begin to discuss properties of fluid flow under ideal conditions. These conditions are that the fluid is incompressible, the temperature is constant, the flow is steady and smooth (or nonturbulent), and there is no viscocity. Steady flow means that, at any point, the velocity and pressure remain constant in time. Smooth flow means that if we track the motion of a small region of the fluid, its path will be smooth and well defined. This path is called a *streamline*. Some of the effects of viscosity and turbulence will be briefly discussed in the next section.

  During the steady flow of a constrained fluid (e.g., through a pipe), the same amount of mass passes through each cross section of pipe in a given amount of time. This steady-flow condition leads to what is known as the **equation of continuity**, which says that the mass m_1 flowing through an area A_1 in a given time equals the mass m_2 flowing through area A_2 in that same amount of time. The amount of mass per unit time of a fluid of density ρ flowing through an area A at speed v is $\rho v A$. Therefore, the equation of continuity is

$$\rho_1 v_1 A_1 = \rho_2 v_2 A_2$$

Under conditions of incompressibility, the densities in the equation of continuity are equal, $\rho_1 = \rho_2$, and we can write the equation as

$$A_1 v_1 = A_2 v_2$$

The quantity Av equals the *volume flow rate* of the fluid and is also called the *flux*, Φ. Thus, the preceding equation says that the volume flow rate (or the flux) is constant for an incompressible fluid.

In general, the same concepts that we use to describe the dynamics of particles also apply to fluid dynamics. With fluids it is often more convenient to express these concepts in terms of density and pressure rather than mass and force. One such example comes from the application of the work-energy theorem to fluids. The result is a mathematical relation known as **Bernoulli's equation**.

With a fluid we can replace the concept of a particle with a small region of the fluid called a *fluid element* of density ρ moving at speed v while sweeping out a volume ΔV under the action of a differential fluid pressure Δp. Bernoulli's equation can be obtained by applying the work-energy theorem ($W_{net} = \Delta K$) to this fluid element. The work done on a fluid element due to a change in the fluid pressure is $\Delta W_{pressure} = (p_1 - p_2)(\Delta V)$. The work done by gravity as the fluid element changes vertical level from a height h_1 to h_2 is $\Delta W_{gravity} = -\rho \Delta V (h_2 - h_1)$. The change in the kinetic energy of the fluid element can be written as $\Delta K = \left(\frac{1}{2}\rho v_2^2 - \frac{1}{2}\rho v_1^2 \right)\Delta V$. These three quantities combine to give Bernoulli's equation along a streamline

$$p_1 + \tfrac{1}{2}\rho v_1^2 + \rho g h_1 = p_2 + \tfrac{1}{2}\rho v_2^2 + \rho g h_2$$

This expression holds in the absence of frictional losses.

Another way of stating Bernoulli's equation is

$$p + \tfrac{1}{2}\rho v^2 + \rho g y = \text{constant}$$

This form of the equation helps make the physical consequences a little more clear because you can see, for example, that for a fluid flowing at a constant vertical level, an increase in the speed of flow must be accompanied by a decrease in pressure (and vice versa). This effect, often called *Bernoulli's principle* (or the *Bernoulli effect*), is important in understanding the consequences of airflow in many applications.

You should also notice that Bernoulli's equation is consistent with the dependence of pressure on depth that was discussed above. Examining this dependence leads to what is often called *Torricelli's law* for the speed of a fluid flowing from an aperture in a container placed a depth h below the surface of the fluid. If, for example, both the surface of the fluid and the aperture are open to the air, then the pressure at both locations is atmospheric pressure, so that $p_1 = p_2$ in Bernoulli's equation. Assuming that the fluid is essentially static at the surface ($v_1 = 0$) we find

$$v_2 = \sqrt{2gh}$$

where $h = y_1 - y_2$.

Example 16–4 Water flows in a horizontal segment of pipe at a pressure of 85 kPa with a speed of 2.6 m/s. The pipe widens, so that its area becomes larger by 35%. If the flow is to be at constant pressure, how far above the initial horizontal level should the pipe divert the water?

Setting it up:

The diagram shows a pipe carrying water first horizontally, then uphill, and horizontally again.

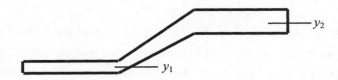

The information provided in the problem is the following:

Given: $p_1 = p_2 = 85$ kPa, $v_1 = 2.6$ m/s, $A_2 = 1.35A_1$;   **Find**: h

Strategy:

Bernoulli's equation relates the pressure to the speed of flow and the change in vertical level. However, the changing width of the pipe can be handled by the equation of continuity, so both expressions should be used.

Working it out:

Using the fact that the pressure is constant, $p_1 = p_2$, Bernoulli's equation becomes,

$$\frac{1}{2}\rho v_1^2 + \rho g y_1 = \frac{1}{2}\rho v_2^2 + \rho g y_2$$

Solving this for the difference in vertical level, we get

$$(y_2 - y_1) = h = \frac{1}{2g}\left(v_1^2 - v_2^2\right).$$

Using the continuity equation to solve for v_2 in terms of v_1 gives

$$v_2 = \frac{A_1 v_1}{A_2} = \frac{A_1 v_1}{1.35 A_1} = \frac{v_1}{1.35}$$

Substituting v_2 into the expression for the change in vertical level produces

$$h = \frac{1}{2g}\left[v_1^2 - \left(\frac{v_1}{1.35}\right)^2\right] = (0.4513)\frac{v_1^2}{2g}.$$

This expression then gives

$$h = (0.4513)\frac{(2.6 \text{ m/s})^2}{2(9.8 \text{ m/s}^2)} = 0.16 \text{ m}.$$

What do you think? If the cross sectional area of the pipe remained constant, what would be the effect on the fluid?

Practice Quiz

7. If the area through which an incompressible fluid flows decreases, the speed of flow will

   (a) decrease   **(b)** increase   **(c)** stay the same

8. If the area through which an incompressible fluid flows is cut in half, the speed of flow will...
   (a) also be cut in half
   (b) double
   (c) decrease to ¼ its speed
   (d) triple
   (e) none of the above

9. If the speed of a horizontally flowing fluid decreases, the pressure in the fluid will
   (a) increase   (b) decrease   (c) stay the same

10. If, for a constant speed of flow, a fluid begins to flow downhill, the pressure in the fluid begins to
    (a) increase   (b) decrease   (c) stay the same

*16–9 Real Fluids

When a particle moves, it usually experiences some sort of frictional resistance to its motion. The same is true for fluid flow. With fluids, this resistance is called **viscosity**. All of the previous discussion assumed an ideal fluid, which, in part, means no viscosity. In this section, we will take a glimpse at the effect of including this unavoidable phenomenon.

To slide a plate along the surface of a fluid at constant velocity requires the application of a force. If the fluid is nonviscous, the force would cause the plate to accelerate, but because of viscosity we require an external force just to keep the velocity constant. Since there is no acceleration, this applied force matches, and therefore characterizes, the viscous force. For a plate of area A sliding at speed v over a fluid film of thickness y, this force is given by

$$F = \frac{\eta v A}{y},$$

where η is called the *coefficient of viscosity*. We can see from this equation that larger values of η mean the fluid is more viscous because greater force is needed for the plate to move at a certain speed. The SI unit of η is N·s/m²; however, a commonly used unit of η is the *poise*, which is defined as

$$1 \text{ poise} = 1 \text{ dyne} \cdot \text{s/cm}^2 = 0.1 \text{ N} \cdot \text{s/m}^2.$$

When the steady flow condition no longer holds, we say that the flow is **turbulent**. The onset of turbulence is characterized by a dimensionless number called the **Reynolds number**, Re. For fluid flow through a tube of length L this number is determined by the equation

$$Re = vL\frac{\rho}{\eta}.$$

Typically, flow becomes turbulent for $Re \gtrsim 2500$.

Practice Quiz

11. In general, we expect the volume flow rate of a fluid with a large coefficient of viscosity to be
    (a) large   (b) small   (c) η has no relevance to the volume flow rate

Reference Tools and Resources

I. Key Terms and Phrases

fluid a liquid or a gas

specific gravity the ratio of the density of a substance to the density of liquid water

buoyancy the phenomenon that fluid pressure applies an upward force on immersed objects

Archimedes' principle that the buoyant force equals the weight of the fluid displaced by an immersed object

equation of continuity the equation that expresses the conservation of mass for a moving fluid

Bernoulli's equation the equation that expresses the conservation of energy for a moving fluid

viscosity resistance to fluid flow

Reynolds number a dimensionless parameter used to determine the onset of turbulence

II. Important Equations

Name/Topic	Equation	Explanation
Pressure	$p = p_0 + \rho g y$	The dependence of pressure on the depth y in a fluid; p_0 is any pressure externally applied to the fluid.
Archimedes' principle	$F_{buoy} = \rho_{fluid} g V_{submerged}$	The buoyant force on a object equals the weight of the fluid displaced by the submerged volume.
Continuity equation	$v_1 A_1 = v_2 A_2$	The volume flow rate in an incompressible fluid is constant.
Bernoulli's equation	$p + \frac{1}{2}\rho v^2 + \rho g h = \text{constant}$	The work-energy theorem applied to fluid flow.
Viscosity	$F = \dfrac{\eta v A}{y}$	The force required to slide an object of cross-sectional area A at speed v over a viscous fluid of depth y.
Turbulence	$Re = v L \left(\dfrac{\rho}{\eta} \right)$	The Reynolds number indicates the onset of turbulence in fluid flow.

III. Know Your Units

Quantity	Dimension	SI Unit
Coefficient of viscosity (η)	$\left[L^{-1} M T^{-1} \right]$	kg/m·s
Reynolds number (Re)	dimensionless	—

Practice Problems

1. In the figure, if $h_1 = 0.12$ cm, $h_2 = 4.41$ cm, and the darker liquid in the U-tube is water (1000 kg/m^3), what is the density of the lighter liquid?

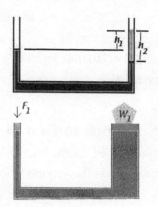

2. In the hydraulic lift at the right the piston on the right has a cross-sectional area of 0.67 m^2, the piston on the left has a cross-sectional area of 0.1 m^2 and the force $F_1 = 194$ N. To the nearest newton, what weight can this configuration lift?

3. On a planet in a different solar system it is known that the sea level atmospheric density, ρ_o, equals 1.45 kg/m^3, sea level pressure, P_o, equals 1.247×10^5 N/m^2, and g equals 9.9 m/s^2. We also know that
$$dP/dy = -P(\rho_o/P_o)g$$
Where g is constant and P varies with altitude y. To the nearest meter, at what altitude is the pressure half of its sea level value?

4. Water is in the big beaker in the figure on the right. Scale 1 reads 99 newtons, scale 2 reads 654 newtons, and scale 3 reads 0 newtons. The hanging block has a density of 10×10^3 kg/m^3. To the nearest tenth of a newton, what does scale 1 read after the block is fully lowered into the beaker of water?

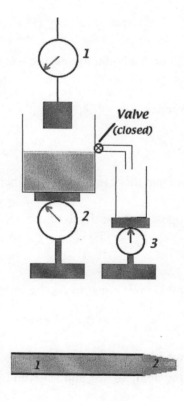

5. In Problem 4, to the nearest tenth of a newton, what is the new reading on scale 2?

6. The experiment in Problem 4 is repeated with the valve opened. What is the new reading on scale 3?

7. In the previous problem, what is the new reading on scale 2?

8. The figure shows a nozzle on a hose carrying water. The pressure at point 1 is 314×10^3 Pa, the velocity is 0.91 m/s, and the diameter is 10 cm. Also, the diameter at point 2 is 2.5 cm. To the nearest tenth of a m/s, what is the velocity at point 2?

9. In Problem 8, to the nearest kPa, what is the pressure at point 2?

10. If for the pipe carrying water in a building, $h =$ 107 m, $v_1 = 2.78$ m/s, and the cross-sectional area at 1 is twice that at 2, what must P_1 be, in order that $P_2 = 101000$ Pa? Give your answer to the nearest kPa.

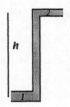

Selected Solutions to End of Chapter Problems

7. The normal forces on the two stilts must equal the weight of the clown. The reaction to the normal force is exerted on the ground, so we have

$$p = N/A = \tfrac{1}{2} \, mg/A = \tfrac{1}{2} \, (68 \text{ kg})(9.8 \text{ m/s}^2) / (4.0 \times 10^{-2} \text{ m})^2 = \boxed{2.1 \times 10^5 \text{ N/m}^2}.$$

19. The force in the large piston must equal the weight of the car. The pressures in the two pistons will be the same, so we have

$$p = F/A_1 = mg/A_2$$

$$\frac{F}{\tfrac{1}{4}\pi D_1^{\,2}} = \frac{mg}{\tfrac{1}{4}\pi D_2^{\,2}}$$

$$\Rightarrow \quad F = mg(D_1/D_2)^2 = (1200 \text{ kg})(9.8 \text{ m/s}^2)[(30 \text{ cm}) / (2 \text{ cm})]^2 = \boxed{52 \text{ N}}.$$

With no frictional losses, the work done by F must increase the potential energy of the car:

$$Fh_1 = mgh_2 \quad \Rightarrow \quad h_2 = Fh_1/mg$$

Therefore,

$$h_2 = \frac{(52 \text{ N})(0.50 \text{ m})}{(1200 \text{ kg})(9.8 \text{ m/s}^2)} = 2.2 \times 10^{-3} \text{ m} = \boxed{2.2 \text{ mm}}$$

Note that this can also be obtained by equating the volume changes.

25. Suppose that the depth of the hull submerged in water is d. Then the volume of the ship submerged in water is $V_{sub} = lwd$, where $l = 400$ m and $w = 60$ m. The corresponding buoyant force from the sea is

$$F_{buoy} = \rho_w V_{sub} \, g = \rho_w (lwd)g.$$

Set this to be equal to F_g, the weight of the ship: $F_g = F_{buoy}$, or $mg = \rho_w (lwd)g$. Thus

$$d = m/(\rho_w lw) = (2.3 \times 10^8 \text{ kg})/[(1.03 \times 10^3 \text{ kg/m}^3)(400 \text{ m})(60 \text{ m})] = \boxed{9.3 \text{ m}} .$$

43. If we ignore air resistance, the water has projectile motion. The maximum range is achieved with an initial angle of 45° and is given by

$$R = v_0^2/g, \quad \Rightarrow \quad v_0 = (gR)^{1/2}.$$

From the mass conservation equation of continuity at the nozzle, with constant density, we have

$$v_1 A_1 = v_2 A_2$$

which becomes

$$(gR_1)^{1/2} \tfrac{1}{4} \pi d_1{}^2 = (gR_2)^{1/2} \tfrac{1}{4} \pi d_2{}^2.$$

After canceling the common factors, we have

$$d_2 = \left(\frac{R_1}{R_2}\right)^{1/4} d_1 = \left(\frac{1.5 \text{ m}}{18 \text{ m}}\right)^{1/4} (1.5 \text{ cm}) = \boxed{0.81 \text{ cm}}$$

53. (a) Because all of the pipelets are equivalent, the speeds must be the same in each.
From the equation of continuity, we have

$$v_0 A_0 = v_1 A_1 + v_2 A_2 + v_3 A_3 + v_4 A_4 = 4 v_1 A_1$$

which gives

$$v_1 = v_2 = v_3 = v_4 = \frac{v_0}{4}\left(\frac{A_0}{A_1}\right) = \frac{v_0}{4}\left(\frac{d_0}{d_1}\right)^2 = \frac{(0.3 \text{ m/s})}{4}\left(\frac{10 \text{ cm}}{2 \text{ cm}}\right)^2 = \boxed{2 \text{ m/s}}$$

(b) If we use Bernoulli's equation between the pipe and a pipelet, we have

$$p_1 + \tfrac{1}{2}\rho v_1{}^2 + \rho g h_1 = p_2 + \tfrac{1}{2}\rho v_2{}^2 + \rho g h_2$$

Solving for p_2 gives

$$p_2 = p_1 + \tfrac{1}{2}\rho\left(v_1^2 - v_2^2\right) - \rho g h_2 = (2.5 \text{ atm})\left(\tfrac{1.01 \times 10^5 \text{ Pa}}{\text{atm}}\right) + \tfrac{1}{2}\left(1.0 \times 10^3 \tfrac{\text{kg}}{\text{m}^3}\right)\left[\left(0.30 \tfrac{\text{m}}{\text{s}}\right)^2 - \left(1.9 \tfrac{\text{m}}{\text{s}}\right)^2\right]$$

$$-\left(1.0 \times 10^3 \tfrac{\text{kg}}{\text{m}^3}\right)\left(9.8 \tfrac{\text{m}}{\text{s}^2}\right)(3.5 \text{ m}) = \boxed{2.2 \times 10^5 \text{ Pa} \ (2.2 \text{ atm})}$$

63. The pressure at a depth h is

$$p = p_0 + \rho g h = (1.01 \times 10^5 \text{ Pa}) + (1.0 \times 10^3 \text{ kg/m}^3)(9.8 \text{ m/s}^2)(30 \text{ m}) = 3.95 \times 10^5 \text{ Pa} = \boxed{3.9 \text{ atm}}.$$

The scuba diver will take in $\boxed{\text{less air}}$, because pressure on the chest and lungs decreases the volume change.

Answers to Practice Quiz

1. (b) **2.** (d) **3.** (c) **4.** (a) **5.** (a) **6.** (e) **7.** (b) **8.** (b) **9.** (a) **10.** (a) **11.** (b)

Answers to Practice Problems

1.	973 kg/m^3	**2.**	1300 N
3.	6021 m	**4.**	89.1 N
5.	663.9 N	**6.**	9.9 N
7.	654.0 N	**8.**	14.6 m/s
9.	208 kPa	**10.**	1162 kPa

CHAPTER 17

TEMPERATURE AND IDEAL GASES

Chapter Objectives

After studying this chapter, you should

1. know how temperature is defined macroscopically.
2. know the relationships between the Kelvin, Celsius, and Fahrenheit temperature scales.
3. be able to account for the thermal expansion of a substance.
4. be able to use the ideal gas law to describe the behavior of ideal gases.
5. know the basic properties of blackbody radiation.

Chapter Review

This is the first of four chapters on **thermodynamics**, which can loosely be described as the study of heat and the physical processes associated with heat transfer. This chapter focuses on the concept of temperature. Most of the key ideas concerning temperature can be conveyed through the study of ideal gases. Apart from being useful to study temperature, ideal gases are important in their own right.

17–1 Temperature and Thermal Equilibrium

In this section, we seek to develop the concepts of *temperature* and **thermal equilibrium**. The two concepts are intimately related and must be defined together. Two objects, or systems, are said to be in *thermal contact* when it is possible for them to exchange energy; however, there may not be a net exchange of energy between systems that are in thermal contact. The property of a system that determines whether or not there will be a net exchange of energy is the temperature. Two systems are said to be in thermal equilibrium if, when they are brought into thermal contact, no net transfer of energy occurs. These statements are embodied in the *zeroth law of thermodynamics*:

> *If system A is in thermal equilibrium with system B, and system C is also in thermal equilibrium with system B, then systems A and C will be in thermal equilibrium if brought into thermal contact.*

In many cases, the system in which we are interested exists within an environment so large by comparison, that the temperature of the environment is virtually unaffected by its interaction with the smaller system. Such a large system is called a thermal reservoir (or heat bath). When a system is in thermal equilibrium with its surrounding thermal reservoir, the temperature of the system, and other temperature-dependent quantities, remain constant. A device that measures the temperature of a system quantitatively is called a **thermometer**.

17–2 Ideal Gases and Absolute Temperature

A low-density (or dilute) gas obeys certain properties that are independent of the type of particles making up the gas (i.e., hydrogen atoms or oxygen molecules, etc.). In particular, for such a gas, the temperature T is directly proportional to the quantity pV/n, where p is the pressure of the gas, V is its volume, and n is the amount of gas present. A gas that obeys this rule is called an **ideal gas**. Because of this behavior, ideal gases can be used as a thermometer.

For an ideal gas at constant volume, then, the gas pressure is directly proportional to the temperature. The temperature that would correspond to a gas of zero pressure is called **absolute zero**. The temperature scale that is established in this way is called the Kelvin scale. Temperature is a new fundamental quantity, in that it is not defined in terms of length, mass, and time; it is assigned the dimensional symbol $[K]$. The resulting relationship between the kelvin temperature and the pressure of a constant-volume ideal gas is

$$T = \left(\frac{T_{tr}}{p_{tr}}\right)p$$

where T_{tr} and p_{tr} are the temperature (273.16 K) and pressure (4.58 mmHg) of the **triple point** of water respectively.

The primary temperature scales in common use are the Celsius, Fahrenheit, and Kelvin scales. Each scale is based on different choices for setting values for two convenient fixed points. The Celsius scale takes the freezing temperature of water to be 0 °C and the boiling temperature of water to be 100 °C. On the Fahrenheit scale, water freezes at 32 °F and boils at 212 °F. In addition to having different settings for the freezing and boiling points of water, the scales of these two systems are different in that a temperature change of one degree on the Fahrenheit scale corresponds to a change of only 5/9 degrees on the Celsius scale. Conversions between the different scales can be accomplished using the following formulas:

$$t_C = \left(1\,°C/K\right)T - 273.15\,°C$$

$$t_F = \left(\frac{9}{5}\,°F/K\right)T - 459.67\,°F$$

$$t_F = \left(\frac{9}{5}\,°F/°C\right)t_C + 32\,°F$$

$$t_C = \left(\frac{5}{9}\,°C/°F\right)\left(t_F - 32\,°F\right)$$

Example 17–1 On a winter day in Michigan, the temperature rose from a morning low of –8.0 °F to an afternoon high of 22 °F. By how many Celsius degrees did the temperature rise?

Setting it up:
The information provided in the problem is the following:
Given: $t_{F,i} = -8.0\,°F$, $t_{F,f} = 22\,°F$;   **Find**: Δt_C in celcius

Strategy/Working it out:
Since we know how to convert from Fahrenheit to Celsius, let's find an expression for ΔT_C.

$$\Delta t_C = t_{C,f} - t_{C,i} = \left(\frac{5}{9}\,°C/°F\right)\left(t_{F,f} - 32\,°F\right) - \left(\frac{5}{9}\,°C/°F\right)\left(t_{F,i} - 32\,°F\right) = \left(\frac{5}{9}\,°C/°F\right)\left(t_{F,f} - t_{F,i}\right)$$

Therefore, we have

$$\Delta T_C = \left(\frac{5}{9}\,°C/°F\right)\left[22\,°F - \left(-8.0\,°F\right)\right] = 17\,C°$$

What do you think?   What is the temperature difference on the Kelvin scale?

Practice Quiz

1. A temperature change of 1 kelvin corresponds to how much of a change in Fahrenheit degrees?
    (a) 0.55   (b) 1.8   (c) 5/9   (d) 32   (e) 273.15

2. What temperature corresponds to absolute zero on the Fahrenheit scale?
    (a) 0 °F   (b) –273.15 °F   (c) – 459.67 °F   (d) – 212 °F   (e) – 100 °F

17–3 Thermal Expansion

Most substances expand when heated and contract when cooled. The amount of expansion is different for different substances. We can identify some aspects of this behavior that are the same for nearly all substances. It is found that the amount by which a substance will expand, that is, its change in length (dL), area (dA), or volume (dV), is directly proportional to the temperature change (dT) that drives the expansion. Furthermore, these changes in size are also directly proportional to the original size (L, A, or V) of the object being heated or cooled.

Thus, for the linear expansion of an object we have $dL \propto L\,dT$. The constant of proportionality depends on the substance and is called the *coefficient of thermal expansion*, α. An expression that describes the linear expansion of an object due to a change in its temperature is

$$\alpha\,\Delta T = \frac{\Delta L}{L}$$

As you can determine by dimensional analysis, the SI unit of α is K^{-1}. Values of the coefficient of linear expansion for a few different substances can be found in Table 17–3 on page 497 of the text.

For the expansion of areas and volumes, the description is very much like that of linear expansion. To a very high approximation (exact for infinitesimally small changes in size), the coefficient of area expansion works out to be twice that of the coefficient of linear expansion, and the coefficient of volume expansion, β, is three times ($\beta = 3\alpha$). This means that we can describe these two phenomena by the expressions

$$(2\alpha)\Delta T = \frac{\Delta A}{A}$$
$$\beta\,\Delta T = \frac{\Delta V}{V}$$

Example 17–2 An iron cube has an edge length of 2.300 cm. If its temperature is raised from –12.00 °C to 22.00 °C, what is its volume at the higher temperature?

Setting it up:

The information provided in the problem is the following:

Given: $l_0 = 2.300$ cm, $t_i = -12.00$ °C, $t_f = 22.00$ °C;   **Find:** V_f

Strategy:

We need to work with the expression for the thermal expansion of volumes. We know that the volume of a cube is given by the cube of its edge length, l^3. We also note that temperature *differences* on the Celcius scale are the same on the Kelvin scale $\Delta t_C = \Delta T$. So, we apply the relation that describes volume expansion to determine ΔV, then add this change to the original volume to get the final result.

Working it out:

Our expression for the change in volume is,

$$\Delta V = \beta_{iron} V_0 (\Delta T),$$

so that

$$V_f = V_0 + \beta_{iron} V_0 (\Delta T) = V_0 \left(1 + \beta_{iron}\, \Delta T\right) = l_0^3 \left(1 + \beta_{iron}\, \Delta T\right)$$

From Table 17–3 in the textbook (p. 497) we are given the value of α_{iron}; so we use $\beta = 3\alpha$. The change in temperature is $\Delta t_C = 22.00$ °C $-$ $(-12.00$ °C$) = 34.00$ °C. Therefore, we know that $\Delta T = 34.00$ K. So,

$$V_f = l_0^3 \left(1 + \beta\, \Delta T\right) = \left[(2.300 \text{ cm})^3\right]\left[1 + 3\left(19.00 \times 10^{-6} \text{K}^{-1}\right)(34.00 \text{ K})\right] = 12.19 \text{ cm}^3.$$

Here, I used 4 significant digits in the value of α in order to show a difference between the initial and final volumes.

What do you think?   Would you expect a larger or smaller final volume if the cube were made of steel?

Practice Quiz

3. What is the coefficient of volume expansion for copper?
   (a) 29×10^{-6} K^{-1} **(b)** 50×10^{-6} K^{-1} **(c)** 87×10^{-6} K^{-1} **(d)** 44×10^{-6} K^{-1} **(e)** 99×10^{-6} K^{-1}

4. If a 12.6-cm copper rod is cooled from 53.2 °C to –10.8 °C, by how much will its length change?
   (a) 1.3×10^{-2} cm   **(b)** 9.1×10^{-3} cm   **(c)** 1.1×10^{-2} cm   **(d)** 2.3×10^{-3} cm   **(e)** 1.7×10^{-5} cm

17–4 The Equation of State of Gases

In section 17–2 we noted that an ideal gas is characterized by the fact that for a fixed amount of gas at constant volume, $pV/n \propto T$ which implies that $p \propto T$; this fact is often referred to as Gay-Lussac's law. It is also empirically known that for fixed n at constant temperature (**isothermal**), $p \propto 1/V$; this fact is called Boyle's law. Finally, for fixed n at constant pressure (**isobaric**), $V \propto T$, which is called Charles' law. Putting all these facts together, we have the relation $pV \propto nT$. Most commonly, n is the number of moles (mol) of gas particles, where 1 mol of a substance contains 6.022×10^{23} entities; this value is called *Avogadro's number*, N_A. The proportionality constant is denoted by R and is called the universal gas constant. The value of this constant is

$$R = 8.314 \text{ J/(mol} \cdot \text{K)}$$

With this constant, we have what is called the ideal gas law,

$$PV = nRT$$

The ideal gas law can be written in terms of the total number of gas particles N by noting that $N = nN_A$. Making this substitution gives $pV = N(R/N_A)T$. The ratio R/N_A is another constant called Boltzmann's constant, k. Thus, we have

$$PV = NkT$$

as an alternative form of the ideal gas law. The value of Boltzmann's constant is $k = 1.381 \times 10^{-23} \text{ J/K}$.

The state of a system is determined by specifying the values of certain quantities that characterize its state, called state variables. An equation that shows how these quantities depend on one another is called an *equation of state*. For the particles making up a gas, the state variables are the pressure p, temperature T, volume V, and amount N (or n) of the gas. The ideal gas law is the equation of state for an ideal gas.

Example 17–3 An automotive worker needs to pump up an empty tire that has an inner volume of 0.0192 m^3. If the temperature in the manufacturing plant is 28.5 °C, what will be the pressure in the tire, as read on a guage, in psi, if 2.70 moles of air is pumped into it?

Setting it up:
The information provided in the problem is the following:
Given: $V = 0.0192$ m^3, $T = 28.5$ °C, $n = 2.70$ mol;   **Find**: p in psi

Strategy:
We can solve this problem by using the ideal gas law. Since we are given the number of moles of air, we shall use the form that contains n. In SI units, this will give us the pressure in pascals.

Working it out:
Solving the ideal gas law for the total pressure we obtain

$$p_{\text{tot}} = \frac{nRT}{V}$$

Since this expression assumes that T is in Kelvin, we must remember to apply the conversion. Doing this conversion gives a pressure of

$$p_{\text{tot}} = \frac{(2.70 \text{ mol})(8.314 \text{ J/mol} \cdot \text{K})\left[\left(28.5\,^{\circ}\text{C}(\tfrac{\text{K}}{^{\circ}\text{C}}) + 273.15 \text{K}\right)\right]}{0.0192 \text{ m}^3} = 3.527 \times 10^5 \text{ Pa}$$

The above result gives the total pressure in the tire. However, a pressure gauge only reports the pressure above atmospheric pressure (often called the *gauge pressure*); therefore, we need to subtract atmospheric pressure. Thus,

$$p = p_{tot} - p_{atm} = 3.527 \times 10^5 \text{ Pa} - 1.013 \times 10^5 \text{ Pa} = 2.514 \times 10^5 \text{ Pa} \left(\frac{1.450 \times 10^{-4} \text{ psi}}{1 \text{ Pa}} \right) = 36.5 \text{ psi}$$

which gives us the final result in pounds per square inch.

What do you think?   Which unit do you think is more convenient to work with for air pressures in tires?

Practice Quiz

5.   For an ideal gas confined to a constant volume, if the temperature is increased by a factor of 2, what happens to the pressure?

   (a) The pressure becomes a factor of 2 smaller.

   (b) The pressure becomes a factor of 2 larger.

   (c) There is no change in the pressure.

   (d) The pressure goes to zero.

   (e) None of the above.

6.   How many molecules are contained in 5.70 moles of a gas?

   (a) 6.02×10^{23} **(b)** 6   **(c)** 3.43×10^{24} **(d)** 22   **(e)** none of the above

*17–5 Blackbody Radiation

All objects emit and absorb heat in the form of *radiation*. By radiation we mean electromagnetic waves, which will be studied in a later chapter. Visible light, infrared, microwaves, X-rays, gamma rays, radio and television waves, and ultraviolet light are all forms of electromagnetic radiation. Infrared and visible light are the two that are most applicable to everyday objects. An object that absorbs all radiation incident upon it is called a blackbody. Note that the ability to establish thermal equilibrium requires that blackbodies emit radiation as efficiently as they absorb it. The main characteristic of **blackbody radiation** is that it obeys the Planck formula

$$u(f, T) = \frac{8\pi h}{c^3} \frac{f^3}{e^{hf/kT} - 1}$$

where f is the frequency of the radiation, h is Planck's constant ($h = 6.625 \times 10^{-34} \text{ J} \cdot \text{s}$), c is the speed of light ($c = 3 \times 10^8 \text{ m/s}$), and the function $u(f,T)$ is defined such that $u(f,T)df$ is the energy density of radiation emitted within the frequency range from f to $f + df$ when the blackbody is at temperature T. One consequence of the Planck formula is that the total power per unit area radiated from the surface of a blackbody (including all frequencies) is proportional to the fourth power of the temperature

$$E(T) = \sigma T^4 .$$

This result is called the *Stefan-Boltzmann formula*; the constant is $\sigma = 5.67 \times 10^{-8} \text{ W/} \left(\text{m}^2 \cdot \text{K}^4 \right)$.

Example 17–4 A carbon ball has a surface area 0.66 m^2 and is heated to a temperature of 92 °C and placed in a room in which the air temperature is 22 °C. If we approximate this ball as a blackbody, at what net rate does radiant heat flow between the ball and the air?

Setting it up:

The information provided in the problem is the following:

Given: $t_{ball} = 92$ °C, $t_{air} = 22$ °C, $A = 0.66$ m^2;   **Find**: P_{net}

Strategy:

Because blackbodies emit radiation as efficiently as they absorb it, the Stefan-Boltzmann formula can be used for both the absorption and the emission. The rate of emission is determined by the temperature of the ball, while the rate of absorption is determined by the temperature of the surrounding air.

Working it out:

The rate at which the ball radiates energy away is given by

$$P_{out} = E_{out} A = \sigma A T_{ball}^4$$

The rate at which the ball absorbs energy from the air is given by

$$P_{in} = E_{in} A = \sigma A T_{air}^4$$

Therefore, the net rate at which the ball loses energy is

$$P_{net} = \sigma A \left(T_{ball}^4 - T_{air}^4 \right).$$

Before using this result for P_{net}, let's first convert the temperatures to Kelvin:

$$T_{ball,K} = t_{ball} + 273.15 = 92 + 273.15 = 365.15 \text{ K}$$
$$T_{air,K} = t_{air} + 273.15 = 22 + 273.15 = 295.15 \text{ K}$$

We can now calculate the net power radiated by the ball as

$$P_{net} = \left(5.67 \times 10^{-8} \ \frac{\text{W}}{\text{m}^2 \cdot \text{K}^4} \right) \left(0.66 \text{ m}^2 \right) \left[\left(365.15 \text{ K} \right)^4 - \left(295.15 \text{ K} \right)^4 \right] = 380 \text{ W}$$

What do you think? How would things be different if the surrounding air were at a higher temperature than the carbon ball?

Practice Quiz

7. At what rate does a blackbody absorb radiation if it has a surface area of 0.33 m^2 and sits in an environment of ambient temperature 290 K?

    (a) 130 J/s   **(b)** 3.0×10^{-6} J/s   **(c)** 290 J/s   **(d)** insufficient information to determine

Reference Tools and Resources

I. Key Terms and Phrases

thermodynamics the branch of physics that studies thermal phenomena on a macroscopic scale

thermal equilibrium exists when systems are brought into thermal contact, and no heat transfer occurs

thermometer a device that measures the temperature of a thermal system

ideal gas a gas in which the gas particles do not interact except through elastic collisions

absolute zero the temperature at which the internal pressure of an ideal gas goes to zero

triple point the state at which a substance can coexist in the solid, liquid, and gaseous forms

mole the amount of a substance that contains 6.022×10^{23} entities

isothermal transformation a thermal process that takes place at constant temperature

isobaric transformation a thermal process that takes place at constant pressure

blackbody radiation the radiation emitted by an object that perfectly absorbs all radiation striking its surface

II. Important Equations

Name/Topic	Equation	Explanation
Temperature scales	$t_C = T - 273.15$ $t_F = \frac{9}{5}T - 459.67$	How the Celsius and Fahrenheit temperatures depend on the Kelvin temperature.
Thermal expansion	$\alpha = \frac{1}{L}\frac{dL}{dT}$ $\beta = \frac{1}{V}\frac{dV}{dT}$	The behavior of materials for linear (upper) and volume (lower) thermal expansion.
Ideal gases	$pV = nRT = NkT$	The relationship (equation of state) between the state variables of pressure, volume, and temperature of an ideal gas.
Blackbody radiation	$E(T) = \sigma T^4$	The power per unit area radiated by a black body as a function of temperature.

III. Know Your Units

Quantity	Dimension	SI Unit
Temperature (T)	$[K]$	K
Coefficient of thermal expansion (α, β)	$\left[K^{-1}\right]$	K^{-1}

Mole (n)	dimensionless	mol
Universal gas constant (R)	$\left[ML^2T^{-2}K^{-1} \right]$	J/(mol·K)
Boltzmann's constant (k)	$\left[ML^2T^{-2}K^{-1} \right]$	J/K
Stefan-Boltzmann constant (σ)	$\left[MT^{-3}K^{-4} \right]$	W/(m²·K⁴)

Practice Problems

1. You inflate your car tires to a gauge pressure of 1.72 atmospheres at a temperature of 4 °C. After driving a couple of miles, the temperature of the tire increases by 27 °C. What is the new gauge pressure, to the nearest hundredth of an atmosphere?

2. To three decimal places, what fraction of the air would you have to let out of the tire in order for the pressure to return to its original value?

3. A 1-cm³ air bubble, at a depth of 136 m and at a temperature of 4 °C, rises to the surface of a lake where the temperature is 16.3 °C. To the nearest tenth of a cm³, what is its new volume?

4. To 1 decimal place, how many micromoles of air is in the bubble in the preceding question?

5. If the temperature in Celsius is 11 , what is the temperature in kelvin, to the nearest degree?

6. If the temperature in Celsius is 118 , what is the temperature in Fahrenheit, to the nearest degree?

7. Concrete ($\alpha = 12 \times 10^{-6}$ K⁻¹) sidewalk slabs 7.1-meters long at 0 degrees Celsius are to be exposed to a temperature range from 0 to 59 degrees Celsius. To the nearest tenth of a mm, how wide should the cracks be?

8. To the nearest kilogram, what is the mass of the air in a room 10 m × 10 m × 3 m, if the air temperature is 45 degrees Celsius and atmospheric pressure is 1×10^5 N/m²? (One mol of air has a mass of 0.029 kg.)

9. A simple pendulum, consisting of a mass on a thin aluminum rod, has a period of 1 second at 10 °C. To the nearest microsecond, by how much is its period increased at 20 °C?

10. One surface is 8 times as hot (emits that much more radiant power) as a second surface. If the second surface has a temperature of 278 K, to the nearest kelvin, what is the temperature of the first surface?

Selected Solutions to End of Chapter Problems

11. (a) For the Celsius scale, we use
$$t_C = T - 273 = 730\text{ K} - 273 = \boxed{457\ °C}.$$
For the Fahrenheit scale, we use
$$t_F = (9/5)T - 460 = (9/5)(730\text{ K}) - 460 = \boxed{854\ °F}.$$
(b) For the Celsius scale, we use
$$t_C = T - 273 = 77\text{ K} - 273 = \boxed{-196\ °C}.$$
For the Fahrenheit scale, we use
$$t_F = (9/5)T - 460 = (9/5)(77\text{ K}) - 460 = \boxed{-321°F}.$$
(c) For the Kelvin scale, we use
$$t_F = (9/5)T - 460 \quad \Rightarrow \quad 75°F = (9/5)T - 460, \text{ which gives } T = \boxed{297\text{ K}}.$$
For the Celsius scale, we use
$$t_F = (9/5)t_C + 32 \quad \Rightarrow \quad 75°F = (9/5)t_C + 32, \text{ which gives } t_C = \boxed{24\ °C}.$$
(d) For the Celsius scale, we use
$$t_F = (9/5)t_C + 32 \quad \Rightarrow \quad 36\ °F = (9/5)t_C + 32, \text{ which gives } t_C = \boxed{2.2\ °C}.$$
For the Kelvin scale, we use
$$t_F = (9/5)T - 460 \quad \Rightarrow \quad 36°F = (9/5)T - 460, \text{ which gives } T = \boxed{276\text{ K}}.$$

23. Both the water and the copper will expand. The copper expansion will increase the volume of the bowl, so for the excess that spills we have
$$\Delta V_w - \Delta V_c = V_0(\beta_w - \beta_c)\,\Delta T$$
$$= (1500\text{ cm}^3)(2.07 \times 10^{-4}\text{ K}^{-1} - 5.10 \times 10^{-5}\text{ K}^{-1})(50\ °C - 20\ °C) = \boxed{7.0\text{ cm}^3}.$$
Note that a temperature change has the same magnitude in °C and K.

45. (a) The pressure is
$$p = (0.60\text{ atm})(1.01 \times 10^5\text{ Pa/atm}) = \boxed{6.1 \times 10^4\text{ Pa}}.$$
The temperature is
$$T = 35°C + 273 = \boxed{308\text{ K}}.$$
(b) The volume of the container is the volume of the gas:
$$pV = NkT \quad \Rightarrow \quad V = NkT/p$$
Therefore,
$$V = \frac{(5.0\times10^{22})(1.38\times10^{-23}\text{J/K})(308\text{ K})}{6.06\times10^4\text{ Pa}} = 3.5\times10^{-3}\text{m}^3$$
(c) Because the volume is constant, we have
$$p_2/p_1 = T_2/T_1 \quad \Rightarrow \quad p_2/(0.60\text{ atm}) = (120 + 273)\text{K}/308\text{ K},$$

which gives $p_2 = \boxed{0.77 \text{ atm}}$.

49. From $pV = nRT$, we have

$$V = nRT/p = (1 \text{ mol})(8.314 \text{ J/mol·K})(273 \text{ K})/(1.01 \times 10^5 \text{ Pa}) = \boxed{22.4 \times 10^{-3} \text{ m}^3}.$$

Given the mass of 1 mol, we find the mass density from

$$\rho = m/V = (28.9 \times 10^{-3} \text{ kg})/(22.4 \times 10^{-3} \text{ m}^3) = \boxed{1.29 \text{ kg/m}^3}.$$

73. (a) $f = kT/h = (1.38 \times 10^{-23} \text{ J/K})(3 \text{ K})/(6.63 \times 10^{-34} \text{ J·s}) = \boxed{6.25 \times 10^{10} \text{ Hz}}$.
(b) $f = kT/h = (1.38 \times 10^{-23} \text{ J/K})(280 \text{ K})/(6.63 \times 10^{-34} \text{ J·s}) = \boxed{5.83 \times 10^{12} \text{ Hz}}$.
(c) $f = kT/h = (1.38 \times 10^{-23} \text{ J/K})(800 \text{ K})/(6.63 \times 10^{-34} \text{ J·s}) = \boxed{1.67 \times 10^{13} \text{ Hz}}$.
(d) $f = kT/h = (1.38 \times 10^{-23} \text{ J/K})(3000 \text{ K})/(6.63 \times 10^{-34} \text{ J·s}) = \boxed{6.25 \times 10^{13} \text{ Hz}}$.

87. The volume and mass of the gas do not change, and the pressure difference is

$$p_2 - p_1 = mg/A$$

where m is the additional mass added to the piston. The two states of the gas are

$$p_1 V = nRT_1 \quad \text{and} \quad p_2 V = nRT_2$$

If we subtract these equations, we get

$$(p_2 - p_1)V = nR(T_2 - T_1),$$
$$V/n = R(T_2 - T_1)/(p_2 - p_1) = R(T_2 - T_1)A/mg$$
$$= (8.314 \text{ J/mol·K})(80 \text{ °C} - 20 \text{ °C})(70 \text{ cm}^2)(10^{-4} \text{ m}^2/\text{cm}^2)/[(0.5 \text{ kg})(9.8 \text{ m/s}^2)]$$
$$= 0.7 \text{ m}^3/\text{mol}.$$

The volume of 0.2 mol is

$$V = (0.7 \text{ m}^3/\text{mol})(0.2 \text{ mol}) = \boxed{0.1 \text{ m}^3}.$$

Answers to Practice Quiz

1. (b) **2**. (c) **3**. (b) **4**. (a) **5**. (b) **6**. (c) **7**. (a)

Answers to Practice Problems

1. 1.98 atm   **2.** 0.089
3. 14.8 cm^3 **4.** 623.9 μmol
5. 284 K   **6.** 244 °F
7. 5.0 mm   **8.** 329 kg
9. 115 μs   **10.** 468 K

CHAPTER 18

HEAT FLOW AND THE FIRST LAW OF THERMODYNAMICS

Chapter Objectives

After studying this chapter, you should

1. understand the meaning and properties of heat flow.
2. be able to do basic calorimetry calculations.
3. be able to calculate the rate of heat flow in materials.
4. be able to calculate the work done by thermal systems.
5. know and be able to use the first law of thermodynamics.
6. be able to determine many properties of an ideal gas including changes in internal energy, heat capacities at constant volume and pressure, and the characteristics of adiabatic transformations.
7. be able to approximate the variation of atmospheric temperature with height.

Chapter Review

In this second chapter on thermodynamics, we discuss the concept of heat and some of its consequences. The discovery of the connection between heat and energy was one of the most important discoveries in the history of science because it affirmed the wide applicability of the law of the conservation of energy, which we now believe to be universally applicable.

18–1 Changes in Thermal Systems

Recall from chapter 17 that the state of a thermodynamic system is characterized by just a few quantities, such as the pressure, temperature, volume, and amount of a gas. If one or more of these quantities change, then there has been a *change of state* of our system. In this chapter, we consider only changes that occur in such a way that the system remains in thermal equilibrium with its surroundings throughout the process that causes the change. When this happens the system is said to have undergone a **reversible process**, which means that both the system and its environment can be returned to their precise states at the beginning of the process. A process for which this condition does not apply is called an **irreversible process**. An example of a reversible process is the slow, controlled expansion of a gas using a piston. An irreversible process would be the free (uncontrolled) expansion of a gas into an open region.

18–2—18–4 Heat Flow, Heat Flow in Materials & The Mechanical Equivalent of Heat

In chapter 17 we noted that when two systems are in thermal contact, there may be an exchange of energy from one system to another. That exchange, or flow, of energy is what we call *heat*. Heat is energy that flows between two systems. To emphasize this fact, we use the term **heat flow**, despite the fact that it is redundant; use of this term also serves to fight the common tendency to say that objects contain a certain amount of heat. The symbol for heat flow is either Q or ΔQ.

Because heat flow is a transfer of energy, and we also know that mechanical work is the transfer of energy, it follows that heat flow can at least sometimes be converted into mechanical work. Two specialized units have been adopted for dealing with the mechanical work associated with heat flow. One unit of heat in common use is the *calorie* (cal). One calorie is the amount of heat needed to raise the temperature of 1 gram of water by 1 Celsius degree. The equivalent amount of energy in joules is called the **mechanical equivalent of heat**:

$$1 \text{ cal} = 4.185 \text{ J}$$

Also in common use is the kilocalorie (kcal), sometimes called a "food calorie" because it is used when quoting the energy content of foods. It is important to realize that the most common notation for kilocalories is *Cal* with a capital *C*, so, whenever energy in calories is being discussed you must note whether or not the *C* is capitalized. Another unit of heat energy in common use is the *British Thermal Unit* (Btu). One Btu is the amount of energy needed to raise the temperature of one pound of water by 1 Fahrenheit degree. In terms of the other units of heat we have

$$1 \text{ Btu} = 252.02 \text{ cal} = 1055 \text{ J}$$

Heat Capacity

One physical consequence of heat flow is a temperature change. The heat flow needed to change the temperature of an *object* is directly proportional to the change in temperature required; that is, $\Delta Q \propto \Delta T$. The proportionality constant between the two quantities, C, is called the **heat capacity** of the object. So,

$$\Delta Q = C \, \Delta T$$

The value of the heat capacity depends on the type of substance as well as the amount (or mass). It is often more useful to have a quantity that is independent of the amount of substance. One such quantity is called the **specific heat** (c), which is basically the heat capacity per unit mass of the substance: $c = C/m$. Another commonly used quantity is the *molar heat capacity* (c'), which is the heat capacity per mole of the substance: $c' = C/n$. Therefore, in terms of these quantities, the amount of heat needed to change the temperature of a *substance* is given by

$$\Delta Q = mc \, \Delta T = nc' \, \Delta T$$

The common unit of specific heat is cal/(g·K) and of molar heat capacity is cal/(mol·K). Table 18–1 on page 520 of the text contains values of these quantities for a few substances.

When heat flow causes a temperature change, the conditions under which this change occurs matter. In general, the heat flow required to obtain a certain temperature change will be different if the volume is held fixed than it will be if the pressure is held fixed. This means that the heat capacity for constant volume processes, C_V, is different from that for processes that take place at constant pressure, C_p. This distinction is most important for gases where pressure and volume can change considerably with temperature. For gases, it is generally the case that $C_p \geq C_V$. The determination of heat capacities is called **calorimetry**.

Example 18–1 An insulated container of negligible heat capacity contains 2.11 kg of water at 22.0 °C. A hot aluminum ball of mass 0.435 kg and temperature 90.0 °C is placed into the water. What will be the final equilibrium temperature of the ball and the water?

Setting it up:

The information provided in the problem is the following:

Given: $m_{Al} = 0.435$ kg, $m_W = 2.11$ kg, $t_{0,W} = 22.0$ °C, $t_{0,Al} = 90.0$ °C;   **Find**: t_f

Strategy

Since the container has negligible heat capacity, we can ignore any heat flow to or from it. Heat will flow from the hot ball to the cooler water. Energy conservation requires that the net heat flow into or out of the system be zero.

Working it out:

By energy conservation we can say that $\Delta Q_W + \Delta Q_{Al} = 0$. The heat flow into the water is

$$\Delta Q_W = m_W c_W \left(t_f - t_{0,W} \right)$$

The heat flow into the aluminum ball is

$$\Delta Q_{Al} = m_{Al} c_{Al} \left(t_f - t_{0,Al} \right)$$

Our energy conservation equation becomes

$$m_W c_W \left(t_f - t_{0,W} \right) + m_{Al} c_{Al} \left(t_f - t_{0,Al} \right) = 0$$

Solving this equation for the final temperature gives

$$t_f = \frac{m_W c_W t_{0,W} + m_{Al} c_{Al} t_{0,Al}}{m_W c_W + m_{Al} c_{Al}}$$

To use this result we must obtain the values of the specific heats from Table 18–1 in the textbook. There, the values are given in cal/g·K. However, the "K" in this unit is for a temperature difference in kelvins; because temperature differences are the same in kelvins as in Celcius degrees, the values of the specific heats are the same if we take the unit as cal/g·°C. Therefore, converting the masses to grams, we have

$$t_f = \frac{(2110 \text{ g})\left(1.00 \frac{\text{cal}}{\text{g·°C}}\right)(22.0\,^\circ\text{C}) + (435 \text{ g})\left(0.22 \frac{\text{cal}}{\text{g·°C}}\right)(90.0\,^\circ\text{C})}{(2110 \text{ g})\left(1.00 \frac{\text{cal}}{\text{g·°C}}\right) + (435 \text{ g})\left(0.22 \frac{\text{cal}}{\text{g·°C}}\right)} = 25\,^\circ\text{C}$$

What do you think? What would be the effect of the container on the final equilibrium temperature if it did not have a negligible heat capacity?

Phase Changes

Generally, heat flow is required to change the phase of a substance. During a phase change, there is no change in the temperature of the system undergoing the transformation, as all the heat flow is used to bring about the phase change. The amount of heat that is required to completely convert one kilogram of a substance from one phase to another is called the **latent heat**, L. Therefore, the amount of heat required to convert a mass m of a substance from one phase to another is given by

$$\Delta Q = mL$$

A common unit of latent heat is cal/g. The value of L depends on the type of phase change being considered. For phase changes between the liquid and gas phases, we use the *latent heat of vaporization*, L_v, and between the liquid and solid phases we use the *latent heat of fusion, L_f.* Some values of L_v and L_f are listed in Table 18–2 on page 522 of the text.

Example 18–2 How much heat is required to convert 2.88 kg of ice at 0.00 °C to steam at 100.0 °C if the mixture is housed in an insulated container?

Setting it up:
The information provided in the problem is the following:
Given: $m = 2.88$ kg, $t_0 = 0.00$ °C, $t_f = 100.0$ °C;;   **Find:** ΔQ_{total}

Strategy:
We can divide the process into three steps. First, the ice melts into liquid water at 0.00 °C, then the liquid warms to the boiling temperature, and then it boils to become steam. In all phases, the mass remains equal to the original mass of the ice. Thus, $\Delta Q_{total} = \Delta Q_{melt} + \Delta Q_{warm} + \Delta Q_{boil}$

Working it out:
The heat flow needed to melt the ice at 0 °C is given by

$$\Delta Q_{melt} = mL_f$$

The heat flow needed to warm the liquid water from to the boiling temperature is

$$\Delta Q_{warm} = mc_W\left(t_f - t_0\right)$$

The heat flow needed to convert the liquid to steam at 100 °C is

$$\Delta Q_{boil} = mL_v$$

Therefore, we have

$$\Delta Q_{total} = mL_f + mc_W\left(t_f - t_0\right) + mL_v$$

This gives

$$\Delta Q_{total} = m\left[L_f + L_v + c_W\left(t_f - t_0\right)\right]$$
$$= (2880 \text{ g})\left[79.6\tfrac{cal}{g} + 540\tfrac{cal}{g} + \left(1.00\tfrac{cal}{g\cdot°C}\right)\left(100.0°C - 0.00°C\right)\right] = 2.07 \times 10^6 \text{ cal}$$

where the values for the latent heats were taken from the text (pp. 521 and 522).

What do you think?   Is this a large amount of energy? Compare it to something familiar to you.

Heat Flow in Materials

In chapter 17, we saw that heat flow can occur by radiation, as with radiation to or from a blackbody. Another process by which heat flow occurs is *conduction*; that is, the heat flows through (or within) the material medium. Heat conduction occurs when heat flows directly through a material because of a temperature difference ΔT across the material. Consider an object in the form of a cylinder of circular cross-sectional area A and length L. If the ends of this cylinder have different temperatures T_2 and T_1 (with $T_2 > T_1$), then heat will conduct through the cylinder until the ends are at the same temperature. The average rate at which heat will flow, $\Delta Q/\Delta t$ (where t is time), is found to depend on three clearly identifiable quantities:

- The heat flow rate is directly proportional to the temperature difference: $\Delta Q/\Delta t \propto \Delta T$ ($\Delta T = T_2 - T_1$).
- The heat flow rate is directly proportional to the area through which it flows: $\Delta Q/\Delta t \propto A$.
- The heat flow rate is inversely proportional to the distance through which it flows: $\Delta Q/\Delta t \propto L^{-1}$.

Combining these considerations, we get $\Delta Q/\Delta t \propto A(\Delta T)/L$. The constant of proportionality κ is called the **thermal conductivity** of the substance. The SI unit of thermal conductivity is W/(m·K). Table 18–3 on page 524 of your text contains values of thermal conductivity for several substances. The final result for the average rate of heat flow by conduction is

$$\frac{\Delta Q}{\Delta t} = \kappa A \frac{\Delta T}{L}$$

An alternative way to characterize an object's properties with regard to thermal conduction is to give its **thermal resistance** (or *R value*). This quantity is defined by

$$R \equiv \frac{L}{\kappa}$$

Therefore, a large R value means that the object is a poor conductor of heat; so, R represents its resistance to heat flow. Written in terms of thermal resistance, the average rate of heat flow through a material is

$$\frac{\Delta Q}{\Delta t} = \frac{A \Delta T}{R}$$

Common units for thermal resistance, such as used for home insulation, is $\text{ft}^2 \cdot \text{h} \cdot {}^\circ\text{F/Btu}$, whereas the SI unit is m²·K/W. The conversion between the two is $1 \text{ ft}^2 \cdot \text{h} \cdot {}^\circ\text{F/Btu} = 0.18 \text{ m}^2 \cdot \text{K/W}$. Objects can be used in combination to obtain a specific R value that may be needed. When objects are placed one after the other they are said to be in series. Placing two objects in series produces an overall, or effective, R value that equals the sum of the individual values

$$\text{In series: } R_{eff} = R_1 + R_2$$

When objects are placed side by side they are said to be in parallel. Placing two objects in parallel produces an effective R value that can be determined from the equation

$$\text{In parallel: } \frac{1}{R_{eff}} = \frac{1}{A_1 + A_2}\left(\frac{A_1}{R_1} + \frac{A_2}{R_2}\right)$$

where A_1 is the cross sectional area of the object with thermal resistance R_1, and similarly for A_2. Notice, that the effective thermal resistance is increased by placing objects in series and decreased by placing them in parallel.

Example 18–3 If a 3.0-ft × 4.0-ft glass window, 3.0-mm thick, has an inner surface temperature of 30.0 °F and an outer surface temperature 20.0 °F, at what rate is energy lost through the window?

Setting it up:

The information provided in the problem is the following:

Given: A = 3.0-ft × 4.0-ft, L = 3.0 mm, t_2 = 30.0 °F, t_1 = 20.0 °F;   **Find:** $\Delta Q/\Delta t$

Strategy:

We need to apply the expression for the rate of thermal conduction. Before we use this expression we should convert quantities to SI units.

Working it out:

To convert the temperatures, we can start with the equation given in chapter 17,

$$t_F = \left(\frac{9}{5}\,^\circ\text{F/K}\right)T - 459.67\,^\circ\text{F}$$

This gives

$$\Delta t_F = \left(\frac{9}{5}\,^\circ\text{F/K}\right)\Delta T \quad \Rightarrow \quad \Delta T = \left(\frac{5}{9}\,\text{K/}^\circ\text{F}\right)\Delta t_F = \left(\frac{5}{9}\,\text{K/}^\circ\text{F}\right)(10.0\,^\circ\text{F}) = 5.556\,\text{K}$$

Converting the surface area gives

$$A = 3.0\,\text{ft} \times 4.0\,\text{ft} = 12.0\,\text{ft}^2\left(\frac{0.3048\,\text{m}}{1\text{ft}}\right)^2 = 1.115\,\text{m}^2$$

We can calculate the rate of heat flow using

$$\frac{\Delta Q}{\Delta t} = \kappa A \frac{\Delta T}{L}$$

where the κ is to be taken from Table 18–3 on page 524 in the textbook. Doing so gives

$$\frac{\Delta Q}{\Delta t} = \left(10.5 \times 10^{-4}\,\frac{\text{kW}}{\text{m} \cdot \text{K}}\right)(1.115\,\text{m}^2)\frac{5.556\,\text{K}}{0.0030\,\text{m}} = 2.17\,\text{kW}$$

What do you think?   What are some things you could do to decrease this high rate of heat loss?

Practice Quiz

1.   Which of the following statements is <u>most</u> accurate?
    (a) If there is a temperature difference between systems A and B, heat will flow from A to B.
    (b) If there is a temperature difference between systems A and B, heat will flow from B to A.
    (c) If there is a temperature difference between two systems, heat will flow from one to the other.

(d) If there is a temperature difference between two systems, heat will flow from the system with higher temperature to the system with lower temperature.

(e) If there is a temperature difference between two systems, heat will flow from the system with higher temperature to the system with lower temperature if they are brought into thermal contact.

2. If you eat 2200 Calories of food per day, what is your energy intake in units of Btu?
   (a) 8.7   (b) 4786   (c) 2200   (d) 8700   (e) 1055

3. If an object made out of a certain substance has a heat capacity C, a similar object made of the same substance but having twice the mass will have a heat capacity of
   (a) $C/2$ (b) C^2 (c) C (d) $2C$ (e) $\sqrt{C}$

4. An object is made out of a certain substance that has a specific heat c. To do calorimetry with a similar object made of the same substance but having twice the mass, we must use a specific heat of
   (a) $c/2$ (b) c^2 (c) c (d) $2c$ (e) $\sqrt{c}$

5. Does it require more or less heat flow to fuse a certain amount of liquid water into ice than to convert the same amount of ice into liquid water? (Assume everything takes place at 0 °C.)
   (a) more   (b) less   (c) the same amount   (d) the answer cannot be determined

6. When water vapor is transformed into liquid water
   (a) the heat flow is out of the vapor.
   (b) the heat flow is into the vapor.
   (c) no heat flow is required.
   (d) the liquid water immediately freezes.
   (e) none of the above.

7. Connecting two identical rods end to end has what effect on the rate of heat flow across the two rods compared with the rate at which heat would conduct across just one of them?
   (a) Heat will conduct at twice the rate for the connected rods.
   (b) Heat will conduct at half the rate for the connected rods.
   (c) Heat will conduct at the same rate for both the connected rods and the single rod.
   (d) There will be no heat conduction for the connected rods.
   (e) None of the above.

18–5—18–6 Work Done by Thermal Systems & The First Law of Thermodynamics

When a gas does work on its surroundings during an expansion, the force that does this work comes from the gas pressure. The amount of work dW performed during an expansion dV is given by $dW = pdV$. The total amount of work done, then, is

$$W = \int_{V_1}^{V_2} pdV$$

This net work done depends on the specific path, in a plot of pressure versus volume, taken by the gas during the process. Geometrically, this integral equals the area under the p vs. V curve. If, instead, the gas is being compressed by its surroundings, then the same integral determines the work done *by the gas*, which in this case would be negative. In a **cyclic transformation**, in which a thermal system follows a path that brings it back to its original state, the net work done by the system during the cycle works out to be the area enclosed by the cyclic path in the p vs. V diagram.

More specifically, we can consider three ideal transformations (processes) that real thermal systems can approximate under certain conditions. One case is *isobaric transformations* (constant-pressure). Because the pressure is constant, it is easy to determine the work done by the system

$$W = p\Delta V$$

Using the ideal gas law, we can see that an isobaric process must behave according to $dT/T = dV/V$ and the heat flow is $dQ = C_p dT$. Another ideal process takes place at constant volume, *isochoric transformations*. Because work is done through a displacement, and there is no displacement if the volume is fixed, no work is done by the system in this case. There is heat flow, however, which is given by $dQ = C_V dT$. Lastly, we consider **adiabatic transformations**. By definition, these transformations take place without the flow of heat; so, $dQ = 0$.

The **internal energy** of a system, U, is the sum of all forms of energy within the system. During an adiabatic process, any work done by the system shows up as an equal change in internal energy. The **first law of thermodynamics** is essentially just an application of the conservation of energy extended to nonadiabatic cases (situations involving heat flow). Any system has a certain amount of internal energy U. Changes in the internal energy result from either heat flow into (positive ΔQ) or out of (negative ΔQ) the system, and/or work done by (positive W) or on (negative W) the system. The mathematical expression of this is

$$\Delta U = -\Delta W + \Delta Q$$

Notice that this expression is consistent with the intuitive notion that the internal energy will increase when either heat is added to the system and/or work is done on the system.

An important distinction between internal energy, heat, and work is that changes in internal energy depend only on the initial and final states of the system (which are determined by pressure, temperature, and volume), and U is therefore called a state variable. Both heat and work depend not only on the states involved but also on the process by which a system is changed from one state to another.

Example 18–4 If 4530 J of work is done on a 0.750-kg piece of copper while its temperature rises from 18.2 °C to 31.2 °C, what is the change in internal energy of the piece of copper? (Assume a constant pressure of 1 atm.)

Setting it up:
The information provided in the problem is the following:
Given: $W = 4530$ J, $m = 0.750$ kg, $t_0 = 18.2$ °C, $t_f = 31.2$ °C;   **Find:** ΔU

Strategy:

Because we are given the amount of work, if we can determine the heat flow ΔQ, we can use the first law of thermodynamics to determine ΔU. The heat flow is that associated with the temperature change.

Working it out:

We can use the specific heat for copper given in the text, to determine ΔQ

$$\Delta Q = mc_{Cu}\Delta T = (750 \text{ g})\left(0.092\tfrac{\text{cal}}{\text{g·K}}\right)(13.0 \text{ K}) = 8.97 \text{ cal}\left(\frac{4.185 \text{ J}}{\text{cal}}\right) = 3754 \text{ J}$$

Since the work is done on the piece of copper, the work done by the copper is $\Delta W = -4530$ J. Therefore, the first law of thermodynamics gives

$$\Delta U = -\Delta W + \Delta Q = -(-4530 \text{ J}) + 3754 \text{ J} = 8300 \text{ J}$$

Note that the final result is rounded to two significant digits because the specific heat of copper is only given to two significant digits in Table 18–1 of the textbook.

What do you think? Suppose instead that no external work was done on the copper, but that the thermal expansion of the copper caused it to do work on its surroundings. Would the change in internal energy be greater or lesser than the above value?

Practice Quiz

8. The first law of thermodynamics is most closely related to
   (a) the conservation of energy
   (b) the conservation of linear momentum
   (c) the conservation of angular momentum
   (d) Newton's second law
   (e) none of the above

9. If 13 J of work is done on a system to remove 9.0 J of heat, what is the change in the internal energy of the system?
   (a) – 4.0 J   **(b)** 22 J   **(c)** – 22 J   **(d)** 4.0 J   **(e)** none of the above

10. The graph shown on the right most likely represents what type of process?
   (a) constant pressure   **(b)** constant volume
   (c) isothermal   **(d)** irreversible
   (e) none of the above

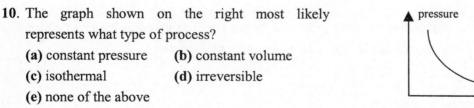

18–7—18–8 Internal Energy of Ideal Gases & More Applications for Ideal Gases

In this section we summarize several key results for the thermal properties of ideal gases.

(a) The internal energy of an ideal gas is a function of temperature only. This implies that, if C_V is independent of temperature (a good approximation under most conditions), then the internal energy is

$$U = C_V T$$

(b) Combining the above result with the first law of thermodynamics and the ideal gas law, we have that the relationship between the heat capacities at constant pressure and volume is

$$C_p = C_V + nR$$

(c) For isothermal transformations, the work done by an ideal is given by

$$W = nRT_0 \ln\left(\frac{V_1}{V_2}\right)$$

where T_0 is the constant temperature of the gas. There is no change in internal energy.

(d) For adiabatic transformations of an ideal gas we have

$$pV^\gamma = \text{constant} \quad \text{and} \quad TV^{\gamma-1} = \text{constant}$$

where γ is the ratio of the heat capacity at constant pressure to that at constant volume

$$\gamma \equiv \frac{C_p}{C_V} = \frac{C_V + nR}{C_V} = \frac{c_V' + R}{c_V'}$$

Example 18–5 A monatomic ideal gas consisting of 6.32 moles of atoms expands from a volume of 14.1 m^3 to 27.6 m^3. How much work is done by the gas if the expansion is **(a)** at a constant pressure of 133 kPa and **(b)** isothermal at $T = 303$ K?

Setting it up:

The information provided in the problem is the following:

Given: $V_1 = 14.1$ m^3, $V_2 = 27.6$ m^3, $n = 6.32$ mol, $p = 133$ kPa, $T_0 = 303$ K

Find: **(a)** W isobaric, **(b)** W isothermal

Strategy:

For part (a) we can use the expression for work done at constant pressure given in the previous section; and for part (b) we can use the expression for work done at constant temperature.

Working it out:

(a) Under conditions of constant pressure,

$$W = p\Delta V = \left(133\times10^3\,\text{Pa}\right)\left(27.6\,\text{m}^3 - 14.1\,\text{m}^3\right) = 1.80\times10^6\,\text{J}$$

(b) The expression for the work done during an isothermal process gives

$$W = nRT_0 \ln\left(\frac{V_2}{V_1}\right) = (6.32 \text{ mol})(8.314 \tfrac{\text{J}}{\text{mol·K}})(303 \text{ K})\ln\left(\frac{27.6 \text{ m}^3}{14.1 \text{ m}^3}\right) = 1.07 \times 10^4 \text{ J}$$

What do you think? How much work would be done by the gas if it were isothermally compressed from 27.6 to 14.1 cubic meters?

Practice Quiz

11. During an isothermal process, a gas expands to twice its previous volume. The pressure in the gas
    (a) doubles
    (b) triples
    (c) becomes larger by a factor of $\sqrt{2}$
    (d) stays the same
    (e) none of the above

Reference Tools and Resources

I. Key Terms and Phrases

reversible process a process that allows a system to return precisely to a previous state

irreversible process a process that is not reversible

internal energy the total of all the forms of energy contained in a system

heat flow the energy that is transferred between systems because of a temperature difference

heat capacity a measure of the heat needed to cause a one degree temperature change of an object

specific heat the heat capacity per unit mass

calorimetry the measurement of heat capacities

latent heat the heat required to cause a phase change

thermal conductivity a quantity that serves as a measure of how readily a substance conducts heat

thermal resistance a measure of an object's resistance to heat flow

mechanical equivalent of heat the conversion between calories and joules

cyclic transformation a transformation in which a thermal system returns to its original state

adiabatic transformation reversible transformations for which there is no heat flow

first law of thermodynamics the conservation of energy for thermal systems

II. Important Equations

Name/Topic	Equation	Explanation
Heat capacity	$\Delta Q = C\Delta T$	The heat required to change the temperature of an object by ΔT.
Latent heat	$\Delta Q = mL$	The heat required to change the phase of an object of mass m.
Heat flow in materials	$\dfrac{dQ}{dt} = -\kappa A \dfrac{dT}{dx}$	The rate at which heat flows through a material of cross sectional area A and thermal conductivity κ.
Thermal resistance	$R = \dfrac{L}{\kappa}$	The thermal resistance of an object of thickness L.
Work done by a gas	$W = \displaystyle\int_{V_1}^{V_2} p\,dV$	The work done by a gas while expanding from volume V_1 to V_2.
First law of thermodynamics	$\Delta U = -W_{A\to B} + Q_{A\to B}$	The statement of the conservation of energy for thermal systems.
Heat capacities of ideal gases	$C_p = C_V + nR$	The relationship between the heat capacities at constant pressure and volume for an ideal gas.
Adiabatic transformations of ideal gases	$PV^\gamma = \text{const}$ $TV^{\gamma-1} = \text{const}$ $\gamma \equiv C_p / C_V$	The equations that describe the behavior of ideal gases during adiabatic transformations.

III. Know Your Units

Quantity	Dimension	SI Unit
Temperature (T)	$[K]$	K
Coefficient of thermal expansion (α, β)	$\left[K^{-1}\right]$	K^{-1}
Mole (n)	dimensionless	mol
Universal gas constant (R)	$\left[ML^2T^{-2}K^{-1}\right]$	J/(mol·K)
Boltzmann's constant (k)	$\left[ML^2T^{-2}K^{-1}\right]$	J/K
Stefan-Boltzmann constant (σ)	$\left[MT^{-3}K^{-4}\right]$	W/(m²·K⁴)

Practice Problems

1. In the figure, 7 mol of an ideal gas goes from point 1, $T_1 = 277$, to point 2, $T_2 = 333$. What is its change in internal energy, to the nearest tenth of a joule?

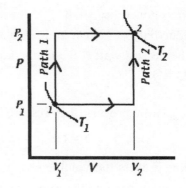

2. If $P_1 = 193$ kPa, $V_1 = 5973$ cm^3 and $P_2 = 200$ kPa, $V_2 = 10000$ cm^3, what is the heat absorbed (+) or liberated (–) to the nearest tenth of a joule in Path 1? Assume the change in internal energy is the same as your answer to Problem 1.

3. If Path 2 was followed in the previous problem, what would the answer be? Assume the change in internal energy is the same as your answer to Problem 1.

4. 10 moles of a gas in thermal contact with an oil bath at a temperature of 300 K is compressed isothermally from a volume of 8874 cm^3 to a volume of 1879 cm^3. To the nearest tenth of a joule, what is the work done by the piston?

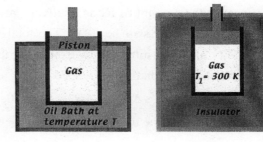

Problem 4   Problem 5

5. What would the answer to the previous problem be if the gas were compressed adiabatically? (It is a monatomic ideal gas.)

6. The heat engine in the figure absorbs $Q_h = 3919$ joules of heat and ejects $Q_c = 1394$ joules of heat. What is its efficiency, to two decimal places?

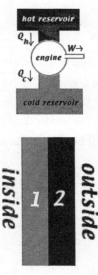

7. A wall is composed of two materials. Material 1 has a thickness of 6.55 cm and a thermal conductivity of 0.1 J/s·m·°C, while material two has a thickness of 4.09 cm and a conductivity of 1 J/s·m·°C. If the temperature difference inside to outside is 25 °C and the wall has an area of 10 m^2, what is the energy loss per second, to the nearest watt?

8. In Problem 7, if the inside temperature is 20.2 °C, what is the temperature of the interface, to the nearest degree Celsius?

9. In the figure (not to scale), a special wall has three metal studs with conductivity of 317 W/m·K. The wall otherwise has a conductivity of 1.66 W/m·K. What is the energy loss per second to the nearest watt if the temperature difference across the wall is 25 °C?

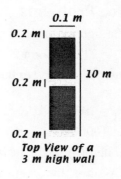

0.1 m

0.2 m|

0.2 m|

10 m

0.2 m|

0.2 m|

Top View of a
3 m high wall

10. In a specific process, the pressure is related to the volume by

$$P = 1.31 \times 10^5 \text{ N/m}^2 + (4.86 \times 10^4 \text{ N/m}^8)V^2$$

In this process the volume changes from 1 m³ to 6.8 m³. What is the work done, to the nearest joule?

Selected Solutions to End of Chapter Problems

7. We neglect the heat capacity of the glass.

   (a) Because the amount of ice is relatively large, we assume that not all of the ice will melt and the final temperature will be 0 °C. The total heat flow into the system is zero. We can check the validity of this assumption by examining the result it produces. If we let m represent the mass in grams of the ice that melts, we have

   $$\Delta Q_{sys} = \Delta Q_w + \Delta Q_i = 0, \quad \text{or} \quad m_w c_w \Delta T_w + mL_f = 0$$

   Solving this for the mass of the ice gives

   $$m = (m_w c_w \Delta T_w)/L_f$$

   Therefore,

   $$m = \frac{(200 \text{ g})(1.00\tfrac{cal}{g \cdot K})(0°C - 25°C)}{79.6 \text{ cal/g}} = 62.8 \text{ g}$$

   Because this is less than 100 g, our assumption is valid.
   The glass contains 262.8 g of water and 37.2 g of ice at 0 °C.

   (b) Because the amount of ice is relatively small, we assume that all of the ice will melt and only water will be present at a temperature above 0°C. The total heat flow into the system is zero, so

   $$\Delta Q_{sys} = \Delta Q_w + \Delta Q_i = 0, \quad \text{or} \quad m_w c_w \Delta T_w + m_i L_f + m_i c_w \Delta T_i = 0$$

   To find the final equilibrium temperature, T, we rewrite this equation as

   $$m_w c_w (T - T_{0,w}) + mL_f + mc_w (T - T_{0,i}) = 0$$

   Solving this for T gives

   $$T = \frac{c_w (m_w T_{0,w} + m T_{0,i}) - mL_f}{(m_w + m)c_w} = \frac{(1.00\tfrac{cal}{g \cdot K})[(250 \text{ g})(25°C) + (50 \text{ g})(0°C)] - (50 \text{ g})(79.6\tfrac{cal}{g})}{(250 \text{ g} + 50 \text{ g})(1.00\tfrac{cal}{g \cdot K})} = 7.6°C$$

   Because this temperature is greater than the melting temperature, 0 °C, our assumption is valid.
   The glass contains 300 g of water at 7.6°C.

19. We find the rate of thermal energy loss from

$$\frac{dQ}{dt} = kA\frac{\Delta T}{\Delta x}$$

To determine the total area, we note that there will be two walls that are 36 ft × 9 ft and another two walls that are 21 ft × 9 ft. So, the total area is

$$A = 2(36 \text{ ft} \times 9 \text{ ft}) + 2(21 \text{ ft} \times 9 \text{ ft}) = 1026 \text{ ft}^2\left(\frac{0.3048 \text{ m}}{\text{ft}}\right)^2 = 95.32 \text{ m}^2$$

Therefore,

$$\frac{dQ}{dt} = \left(6.3 \times 10^{-4}\,\tfrac{kW}{m \cdot K}\right)\left(95.32 \text{ m}^2\right)\frac{(76^\circ\text{F} - 25^\circ\text{F})\left(\tfrac{5^\circ\text{C}}{9^\circ\text{F}}\right)}{(9 \text{ in})\left(\tfrac{0.0254 \text{ m}}{\text{in}}\right)} = \boxed{7.4 \text{ kW}}$$

33. We assume that all of the kinetic energy loss heats the lead. From conservation of energy during the fall, the kinetic energy of the bag just before hitting the ground is equal to the initial potential energy. So, for N drops, we have

$$\Delta Q = W = -\Delta U, \quad \text{or} \quad mc\Delta T = -(0 - Nmgh),$$

which becomes

$$N = c\Delta T/gh = (128 \text{ J/kg·K})(5.0 \text{ K})/(9.8 \text{ m/s}^2)(2.0 \text{ m}) = 32.7 = \boxed{33 \text{ times}}.$$

43. For the ideal gas, we have

$$pV = nRT \quad \Rightarrow \quad V = nRT/p$$

$V_i = (0.5 \text{ mol})(8.314 \text{ J/mol·K})(293 \text{ K})/(1.0 \text{ atm})(1.01 \times 10^5 \text{ Pa/atm}) = \boxed{0.012 \text{ m}^3};$

$V_f = (0.5 \text{ mol})(8.314 \text{ J/mol·K})(293 \text{ K})/(3.6 \text{ atm})(1.01 \times 10^5 \text{ Pa/atm}) = \boxed{0.0034 \text{ m}^3}.$

For the isothermal case we know that

$$W = nRT \ln(V_f/V_i)$$

and that

$$V_f/V_i = p_i/p_f$$

Putting these together, we have

$W = nRT \ln(p_i/p_f) = (0.5 \text{ mol})(8.314 \text{ J/mol·K})(293 \text{ K}) \ln[(1.0 \text{ atm})/(3.6 \text{ atm})] = \boxed{-1.6 \times 10^3 \text{ J}}.$

The negative value means that work is done on the gas to compress it.

51. We can find the internal energy change from the path $A \rightarrow C \rightarrow B$:

$$\Delta U = U_B - U_A = -W_{A \rightarrow C \rightarrow B} + Q_{A \rightarrow C \rightarrow B}$$
$$= -20,000 \text{ cal} + 40,000 \text{ cal} = +20,000 \text{ cal}.$$

(a) For the path $A \rightarrow D \rightarrow B$, the first law of thermodynamics is

$$U_B - U_A = -W_{A \rightarrow D \rightarrow B} + Q_{A \rightarrow D \rightarrow B};$$
$$+20,000 \text{ cal} = -7,000 \text{ cal} + Q_{A \rightarrow D \rightarrow B}, \text{ which gives}$$
$$Q_{A \rightarrow D \rightarrow B} = \boxed{2.7 \times 10^4 \text{ cal}}.$$

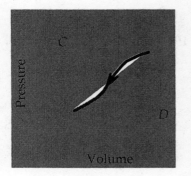

(b) For the return path $B \rightarrow A$, the work done by the gas is negative, and the first law of thermodynamics gives us

$$U_A - U_B = -W_{B \rightarrow A} + Q_{B \rightarrow A};$$
$$-20,000 \text{ cal} = -(-15,000 \text{ cal}) + Q_{B \rightarrow A},$$

which gives $Q_{B \rightarrow A} = \boxed{-3.5 \times 10^4 \text{ cal}}$.

The negative value means the heat is $\boxed{\text{liberated}}$.

57. We find the internal energy change from the first law of thermodynamics:

$$\Delta U = -W + Q = -8.0 \text{ J} + 5.0 \text{ J} = -3.0 \text{ J}.$$

For an ideal gas, we have

$$\Delta U = nc'_V \Delta T$$

$$\Delta T = \frac{\Delta U}{nc'_V} = \frac{-3.0 \text{ J}}{(0.2 \text{ mol})\left(20.8 \frac{\text{J}}{\text{mol} \cdot \text{K}}\right)} = \boxed{-0.72 \text{ K (cooled)}}$$

Note that we use $\Delta U = nc'_V \Delta T$, even though the process is not at constant volume.

Answers to Practice Quiz

1. (e) **2.** (d) **3.** (d) **4.** (c) **5.** (c) **6.** (a) **7.** (b) **8.** (a) **9.** (d) **10.** (c) **11.** (e)

Answers to Practice Problems

1.	4886.3 J	**2.**	4080.9 J
3.	4109.1 J	**4.**	38701.0 J
5.	67867.9 J	**6.**	0.64
7.	359 W	**8.**	−3 °C
9.	154,353 W	**10.**	5,837,398 J

CHAPTER 19

THE MOLECULAR BASIS OF THERMAL PHYSICS

Chapter Objectives

After studying this chapter, you should

1. understand the microscopic origin of pressure.
2. know how the temperature of a gas relates to the average kinetic energy of gas particles.
3. know how to calculate averages from a distribution.
4. have a basic understanding of transport phenomena in gases.

Chapter Review

In this third chapter on thermodynamics, we discuss how thermodynamics can be understood in terms of the microscopic behavior of molecules. The basic theme of this chapter is the kinetic theory of gases.

19–1—19–3 The Microscopic View of Gases

The **kinetic theory** of gases uses average values related to the motion of the microscopic gas particles (atoms or molecules) to explain the macroscopic properties of gases. Specifically, we shall use kinetic theory to gain deeper insight into the origin of the pressure and temperature of a gas. In its simplest form, the kinetic theory of gases makes several reasonable assumptions: (a) we are dealing with a large number, N, of identical gas particles, (b) the particles move randomly, and (c) the only interactions experienced by the particles are elastic collisions with each other (and with the container).

One of the important quantities in understanding the behavior of gases is the average kinetic energy

$$\langle K \rangle = \frac{1}{2} m \langle v^2 \rangle$$

Note that although $\langle \vec{v} \rangle = 0$ because of the randomly directed motion, $\langle v^2 \rangle \neq 0$. Therefore, the useful way to characterize the velocity of a typical gas particle is with the **root-mean-square (rms) speed** given by $v_{rms} = \sqrt{\langle v^2 \rangle}$; this velocity is the velocity of a particle that has the average kinetic energy. The total internal energy of the gas is primarily from the kinetic energy and is equal to the number of particles times the average kinetic energy per particle

$$U = N \langle K \rangle$$

Kinetic theory shows that the pressure exerted by a gas on a container of volume V is due to the collisions of the gas particles with the walls of the container. If you consider the change in momentum, $\Delta \vec{P}$, of a gas particle upon collision with a wall of the container together with the amount of time between collisions with this wall, Δt, the magnitude of the force that this particle exerts on the wall, $F = \Delta P / \Delta t$, can be determined. Upon dividing this force by the area of the wall for all N particles, the pressure works out to be

This latter expression clearly shows that the pressure of a gas is directly proportional to the average translational kinetic energy of the gas particles.

If we compare the above result with the ideal gas law, $pV = NkT$, we find that the Kelvin temperature of a gas is directly proportional to the average kinetic energy of the gas particles. Specifically,

$$\langle K \rangle = \frac{3}{2}kT$$

It is from this result that we gain deep insight into the meaning of temperature: the temperature of an ideal gas is a measure of the average kinetic energy of the constituents. Moreover, this result also shows why the Kelvin temperature scale is a physically more appropriate scale to use in scientific work — it has the most direct connection to the energy of the particles in a system.

Example 19–1 If 15 moles of a gas are contained in a volume of 0.10 m^3 at atmospheric pressure, what is the average kinetic energy of the gas particles?

Setting it up:
The information provided in the problem is the following:
Given: $n = 15$ mol, $V = 0.10$ m^3, $p = 101$ kPa;   **Find**: $\langle K \rangle$

Strategy:
Now that we know how the average kinetic energy fits into the ideal gas law, we can put these results together to determine the answer to this question.

Working it out:
From kinetic theory we have

$$pV = \frac{2}{3}N\langle K \rangle \quad \Rightarrow \quad \langle K \rangle = \frac{3}{2}\frac{pV}{N}$$

The number of particles is related to the number of moles by $N = nN_A$. Thus,

$$\langle K \rangle = \frac{3}{2}\frac{pV}{nN_A} = \frac{3}{2}\frac{\left(101\times10^3 \text{ Pa}\right)\left(0.10 \text{ m}^3\right)}{(15 \text{ mol})\left(6.022\times10^{23} \text{ mol}^{-1}\right)} = 1.7\times10^{-21} \text{ J}$$

What do you think?   What is the temperature of this gas?

Practice Quiz
1. If you pump energy into a gas such that the rms speed of the molecules triples, what happens to the temperature of the gas?
    (a) The temperature of the gas triples.
    (b) The temperature is reduced to one-third its previous value.
    (c) The temperature is reduced to one-ninth its previous value.
    (d) The temperature becomes nine times its previous value.
    (e) None of the above.

2. What is the average kinetic energy of a molecule in a monatomic gas if the temperature of the gas is 20 °C?

(a) 4.1×10^{-22} J   (b) 2.8×10^{-22} J   (c) 4.0×10^{-21} J   (d) 7.3×10^{-21} J   (e) 6.1×10^{-21} J

19–4—19–6 Probability Distributions

As discussed in the previous section, the average value of quantities like speed and kinetic energy are important in a gas. However, the actual values of the speeds and kinetic energies will be distributed in some way around the average. We can statistically characterize these values using *probability distributions*. The probability distribution, $P(x)$, for a quantity x, is a function that contains information on the probability of obtaining a particular value of x. If x is a discrete quantity, such as the grades of students in a physics course, then $P(x)$ is the probability that the value x is realized

$$P(x) = N_x / N \quad \text{(discrete distribution)}$$

where N_x is the total number of x values and N is the total number of values of any kind. We want the total probability of getting one of the possible values to be 1, so we require a normalization condition of

$$\sum_x P(x) = 1$$

Average values in a discrete distribution are determined according to

$$\langle x^n \rangle = \sum_x x^n P(x)$$

For quantities that vary continuously, we can only consider ranges of the quantity because the total number of possibilities is infinite ($N = \infty$) given a probability of zero for obtaining any one value. In this case, the probability distribution, $P(x)$, gives the probability per interval of x, of obtaining a value within that interval. That is

$$P(x)dx = \text{ the probability that } x \text{ lies between } x \text{ and } x + dx$$

The probability of obtaining a value of x within the range x_a to x_b is

$$\text{prob} = \int_{x_a}^{x_b} P(x)dx$$

Similarly, if the entire range of x is from x_1 to x_2, the normalization condition is

$$\int_{x_1}^{x_2} P(x)dx = 1$$

Average values of continuous quantities are calculated according to

$$\langle x^n \rangle = \int_{x_1}^{x_2} x^n P(x)dx$$

Having discussed probability distributions in general, we can now treat specific cases. The first case is the velocity distribution in an ideal gas. The normalized distribution function is given by

$$F(\vec{v}) = \left(\frac{m}{2\pi kT} \right)^{3/2} e^{-mv^2/2kT}$$

Using this distribution function, it can be shown (see your textbook) that

$$v_{rms} = \sqrt{\langle v^2 \rangle} = \sqrt{\frac{3kT}{m}}$$

which is consistent with the results from the previous section.

The second case is the energy distribution in an ideal gas. This distribution, known as the **Maxwell-Boltzmann distribution**, is

$$F(E) = \frac{1}{Z} e^{-E/kT}$$

where the factor Z is there to account for normalization. The energy E in this equation is the total kinetic energy of the gas particles. For molecules, this total energy includes not only the translational kinetic energy, but also rotational and vibrational kinetic energies. Using this distribution to determine the average energy $\langle E \rangle$ leads to the **equipartition theorem**

$$\langle E \rangle = \frac{s}{2} kT$$

where s is the number of independent **degrees of freedom** (ways in which the molecule can move) of the gas constituents. The theorem has this name because it says that the total energy is equally divided between the different degrees of freedom. Notice that because particles, strictly speaking, cannot rotate or vibrate but only move translationally, this theorem reproduces our previous result of $\langle K \rangle = \frac{3}{2} kT$. Having additional degrees of freedom, for molecular gases, also affects the heat capacity of the gas. Using our results, $U = N \langle K \rangle$, and from chapter 18, $U = C_V T$, we can see that

$$C_V = \frac{s}{2} nR$$

Example 19–2 Diatomic nitrogen (N_2) has a *bond length* of 1.10×10^{-10} m (the distance between atomic centers). Treating this molecule in the rigid "dumbbell" approximation, what is its root-mean-square angular velocity if the gas is in equilibrium at a temperature of 315 K?

Setting it up:
By analogy with v_{rms} we conclude that the rms angular velocity is $\omega_{rms} = \sqrt{\langle \omega^2 \rangle}$. This would then represent the angular velocity of the particle with the average rotational kinetic energy $\langle K_{rot} \rangle = \frac{1}{2} I \langle \omega^2 \rangle$. The information provided in the problem is the following:
Given: $l = 1.10 \times 10^{-10}$ m, $T = 315$ K;   **Find**: ω_{rms}

Strategy:
As discussed in this section of your textbook, a dumbbell-like object has five degrees of freedom, the usual three spatial directions plus two additional rotational degrees of freedom. Thus,

$$\langle E \rangle = \tfrac{1}{2} m v_x^2 + \tfrac{1}{2} m v_y^2 + \tfrac{1}{2} m v_z^2 + \tfrac{1}{2} I \omega_x^2 + \tfrac{1}{2} I \omega_y^2 = \tfrac{1}{2} m v_{rms}^2 + \tfrac{1}{2} I \omega_{rms}^2$$

To solve for the factor we want, we'll need to look up the mass of the nitrogen atom.

Working it out:

From Appendix V in your textbook we see that the mass of a single nitrogen atom is 14.00674 u. The conversion factor is given in Appendix I; so,

$$m = 14.00674 \text{ u} \left(\frac{1.661 \times 10^{-27} \text{ kg}}{1 \text{ u}} \right) = 2.3265 \times 10^{-26} \text{ kg}$$

The rotational inertia then is

$$I = m(l/2)^2 + m(l/2)^2 = ml^2/2$$

The equipartition theorem says that the kinetic energy is equally divided between the different degrees of freedom giving an amount of $\frac{1}{2} kT$ to each. Therefore, since the rotational kinetic energy contains two of the five degrees of freedom we have that

$$\tfrac{1}{2} I \omega_{rms}^2 = kT$$

Solving this for the rms angular velocity gives

$$\omega_{rms} = \left(\frac{2kT}{I} \right)^{1/2} = \left(\frac{4kT}{ml^2} \right)^{1/2} = \left[\frac{4(1.381 \times 10^{-23} \text{ J/K})(315 \text{ K})}{(2.3265 \times 10^{-26} \text{kg})(1.10 \times 10^{-10} \text{ m})^2} \right]^{1/2} = 7.86 \times 10^{12} \text{ rad/s}$$

What do you think? This angular velocity seems ridiculously large, but is it? What checks can you make to see if it is a reasonable value for this situation?

Practice Quiz

3. What is the rms speed of the nitrogen gas in Example 19–2?
   (a) 749 m/s   (b) 306 m/s   (c) 432 m/s   (d) 530 m/s   (e) 673 m/s.

4. If the nitrogen gas in Example 19–2 contains 2.44 moles, what is the internal energy of the gas?
   (a) 1.09×10^{-20} J   (b) 1.60×10^4 J   (c) 1.47×10^{24} J   (d) 9.59×10^3 J   (e) 2.65×10^{-20} J

*19–7 Collisions and Transport Phenomena

The molecules in a gas are constantly undergoing collisions. Through these collisions properties of the gas are distributed throughout. The movement of properties through a gas is called *transport*. Closely related to transport is the more general property that gas constituents naturally spread throughout a gas, which is called *diffusion*.

The molecules in a gas collide when they come within a certain distance of each other. As viewed from molecule A, the other molecule, B, presents a certain size, or cross sectional area, that defines its **collision cross section** σ,

$$\sigma = \pi D^2$$

where D is the characteristic distance within which molecule A must pass in order to have a collision with the molecule B. The average amount of time between collisions, called the *mean collision time* τ, certainly depends of σ, but it also depends on how close the molecules are to each other, the number density n (number of molecules per unit volume), and on how fast they are moving v_{rms}. You would expect τ to be short if the molecules are large (large σ), fast moving (large v_{rms}), and closely packed together (large n). The result is

$$\tau = \frac{1}{\sqrt{2}n\sigma v_{rms}}$$

Closely related to the mean collision time is the **mean free path**

$$\lambda = \tau v_{rms} = \frac{1}{\sqrt{2}n\sigma}$$

which is a measure of the average distance a molecule travels between collisions.

As diffusion occurs, the collisions encountered by a particular molecule are random, so it gets knocked around from all directions, and undergoes what is called a **random walk**. This random walk is responsible for *Brownian motion* – the random, jiggling, motion of particles that can be observed in a gas or liquid. During a random walk, the average distance moved in time t can be approximated by

$$\langle r^2 \rangle = \left(\frac{\lambda^2}{\tau} \right) t$$

Practice Quiz

5. Consider two identical containers, A and B, of nitrogen gas held at the same temperature. If container A holds twice as many molecules as container B, which gas molecules will experience a longer mean free path?
    (a) Those in container A
    (b) Those in container B
    (c) They will have equal mean free paths.
    (d) The answer cannot be determined from the above information.

Reference Tools and Resources

I. Key Terms and Phrases

kinetic theory the theory that relates the motion of the microscopic particles making up a system to its macroscopic properties

root-mean-square (rms) speed defined as the square root of $\langle v^2 \rangle$

Maxwell-Boltzmann distribution the function that describes the probability of an ideal gas particle having a certain total energy

degree of freedom a way in which an object can move that is independent of the other ways it can move

equipartition theorem expresses the fact that the energy of a gas particle is equally partitioned between its degrees of freedom

collision cross section the effective area that one molecule presents to the other for a collision to occur

mean free path the average distance a gas particle moves before it is involved in a collision

random walk the motion of a particle that moves in equal steps in random directions

II. Important Equations

Name/Topic	Equation	Explanation
Kinetic theory	$v_{rms} = \sqrt{\langle v^2 \rangle}$	The rms speed is the speed of a gas particle that has the average kinetic energy.
Kinetic theory	$kT = \frac{2}{3}\langle K \rangle$	The kelvin temperature of a gas corresponds to the average kinetic energy of gas particles.
Probability distributions	$\langle x^n \rangle = \int_{x_1}^{x_2} x^n P(x)\,dx$	How to calculate the average of a quantity from a probability distribution.
Maxwell-Boltzmann distribution	$F = \frac{1}{Z} e^{-E/kT}$	The probability distribution for the energy of an ideal gas particle.
Equipartition theorem	$\langle E \rangle = \frac{s}{2} kT$	The relationship between the average energy of a gas molecule and the number of degrees of freedom, s.
Collision cross section	$\sigma = \pi D^2$	The effective area that one molecule presents to the other for a collision to occur.
Mean free path	$\lambda = \dfrac{1}{\sqrt{2}n\sigma}$	The average distance a gas particle moves before it is involved in a collision.

III. Know Your Units

Quantity	Dimension	SI Unit
Maxwell-Boltzmann distribution (F)	$\left[M^{-1} L^{-2} T^2 \right]$	J^{-1}
Collision cross section (σ)	$\left[L^2 \right]$	m^2
Mean free path (λ)	$[L]$	m

Practice Problems

1. What is the most probable speed for molecules ($m = 10 \times 10^{-27}$ kg) if the temperature is 253 kelvin, to the nearest m/s? Note: The Maxwell distribution in terms of speed is proportional to $v^2 e^{-(mv^2/2kT)}$.

2. For the conditions in Problem 1, what is the rms velocity?

3. According to the ideal gas equation, what is the pressure for a gas with a volume of 885 L and a temperature of 298 K if the number of moles is 700? Give your answer to the nearest N/m^2.

4. If the van der Waals equation of state was used and the gas was CO_2, what would be the answer to Problem 3?

5. What is the mean free path, to the nearest tenth of a nanometer, for the molecules in Problem 3? Assume a molecular radius of 1.5×10^{-10} m.

6. What is the average translational kinetic energy in eV (electron-volts) of nitrogen molecules, at a temperature of 132 degrees Celsius? The mass of a nitrogen molecule is 46.5×10^{-27} kg and your answer should be given to three decimal digits.

7. To the nearest m/s, what is the root-mean-square velocity of the molecules in Problem 6?

8. Six molecules have the following speeds (in arbitrary units): 20, 14, 11, 10, 4, 12. To the nearest single decimal digit, what is their rms speed?

9. What is the average speed of the particles in the previous problem?

10. To the nearest tenth of a millielectronvolt, what is the average energy of the H_2 molecule at the temperature of 417 K?

Selected Solutions to End of Chapter Problems

7. **(a)** The internal energy of an ideal gas depends only on the temperature:

$$U = nc'_V T = \tfrac{3}{2} nRT = \tfrac{3}{2} pV = \tfrac{3}{2}(5 \text{ atm})(1.01 \times 10^5 \tfrac{\text{Pa}}{\text{atm}})(5000 \text{ cm}^3)(10^{-6} \tfrac{\text{m}^3}{\text{cm}^3}) = \boxed{3.8 \times 10^3 \text{ J}}.$$

(b) From the ideal gas law, we have

$T = pV/nR$

$= (5 \text{ atm})(1.01 \times 10^5 \text{ Pa/atm})(5000 \text{ cm}^3)(10^{-6} \text{ m}^3/\text{cm}^3)/(1 \text{ mol})(8.314 \text{ J/mol·K}) = \boxed{304 \text{ K}}$.

(c) On the molecular level, the internal energy is the motion of the molecules:

$$U = \tfrac{1}{2} Nm\langle v^2 \rangle \quad \Rightarrow \quad \langle v^2 \rangle = 2U/Nm$$

Therefore,

$$\langle v^2 \rangle = \frac{2(304 \text{ J})}{(6.02 \times 10^{23})(3.36 \times 10^{-26} \text{ kg})} = \boxed{3.8 \times 10^5 \text{ m}^2/\text{s}^2}$$

(d) From part (c), we have

$$v_{\text{rms}} = \langle v^2 \rangle^{1/2} = (3.8 \times 10^5 \text{ m}^2/\text{s}^2)^{1/2} = \boxed{6.1 \times 10^2 \text{ m/s}}$$

17. The average kinetic energy of the molecule is

$$\langle K \rangle = \tfrac{3}{2} kT = \tfrac{3}{2} (1.38 \times 10^{-23} \text{ J/K})(293 \text{ K}) = \boxed{6.1 \times 10^{-21} \text{ J}}.$$

We find the speed of the baseball from

$$K = \tfrac{1}{2} mv^2 \quad \Rightarrow \quad v = \sqrt{2K/m}$$

Therefore,

$$v = \sqrt{\frac{2(6.1 \times 10^{-21} \text{ J})}{0.10 \text{ kg}}} = \boxed{3.5 \times 10^{-10} \text{ m/s}}$$

21. The rms speed depends on the temperature and the mass of a molecule:

$v_{\text{rms}} = (3kT/m)^{1/2}$

$v_{\text{rms,oxygen}} = \{3(1.38 \times 10^{-23} \text{ J/K})(300 \text{ K})/[(32 \times 10^{-3} \text{ kg/mol})/(6.02 \times 10^{23} \text{ molecules/mol})]\}^{1/2}$
$= \boxed{484 \text{ m/s}}$

$v_{\text{rms,nitrogen}} = \{3(1.38 \times 10^{-23} \text{ J/K})(300 \text{ K})/[(28 \times 10^{-3} \text{ kg/mol})/(6.02 \times 10^{23} \text{ molecules/mol})]\}^{1/2}$
$= \boxed{517 \text{ m/s}}$

$v_{\text{rms,carbon dioxide}} = \{3(1.38 \times 10^{-23} \text{ J/K})(300 \text{ K})/[(44 \times 10^{-3} \text{ kg/mol})/(6.02 \times 10^{23} \text{ molecules/mol})]\}^{1/2}$
$= \boxed{412 \text{ m/s}}$

$v_{\text{rms,hydrogen}} = \{3(1.38 \times 10^{-23} \text{ J/K})(300 \text{ K})/[(2.0 \times 10^{-3} \text{ kg/mol})/(6.02 \times 10^{23} \text{ molecules/mol})]\}^{1/2}$
$= \boxed{1934 \text{ m/s}}.$

35. (a) Using the value at the midpoint of each range, we find the average grade from

$\langle g \rangle = (1/N)(\Sigma g N_g)$

$= (1/87)[(-40)(3) + (-20)(18) + (0)(29) + (20)(22) + (40)(15)] = \boxed{+6.4}.$

(b) $\langle D^2 \rangle = \Sigma(D^2 N_D)/N = [\Sigma(g - \langle g \rangle)^2 N_g] / N$

$= (1/87)[(-46.4)^2 (3) + (-26.4)^2 (18) + (-6.4)^2 (29) + (13.6)^2 (22) + (33.6)^2 (15)] = \boxed{474}.$

(c) Because the midpoint of the range is +20, we have

$\langle g \rangle = (1/N)(\Sigma g N_g) = (1/87)(+20)(87) = \boxed{+20}.$

49. (a) The average molecular energy of argon (gas 1), a monatomic gas, is $\tfrac{3}{2} kT$, while that of nitrogen (gas 2), a diatomic gas, is $\tfrac{5}{2} kT$. Take $T \approx 300$ K for room temperature. Then

$$U = U_1 + U_2 = N_1 \langle E_1 \rangle + N_2 \langle E_2 \rangle = n_1 N_A \langle E_1 \rangle + n_2 N_A \langle E_2 \rangle = n_1 N_A (\tfrac{3}{2} kT) + n_2 N_A(\tfrac{5}{2} kT)$$

$$= (3n_1 + 5n_2)RT/2 \approx [3(1.3 \text{ mol}) + 5(2.0 \text{ mol})](8.314 \text{ J/mol·K})(300 \text{ K})/2 = \boxed{17 \text{ kJ}}.$$

(b) The mass of the mixture is $m = (0.040 \text{ kg/mol}) (1.3 \text{ mol}) + (0.028 \text{ kg/mol})(2.0 \text{ mol}) = 0.128 \text{ kg}$. So, the specifc heat of the mixture is

$$c_v = (dU/dT) / m = \frac{1}{m}\frac{d}{dT}[(3n_1 + 5n_2) RT/2] = \frac{R}{2m}(3n_1 + 5n_2)$$

$$= \frac{8.314 \text{ J/mol·K}}{2(0.128 \text{ kg})}[3(1.3 \text{ mol}) + 5(2.0 \text{ mol})] = \boxed{0.45 \text{ kJ/kg·K}}.$$

67. The pressure is

$$p = [(10^{-9} \text{ torr})/(760 \text{ torr/atm})](1.01 \times 10^5 \text{ Pa/atm}) = 1.3 \times 10^{-7} \text{ Pa}.$$

(a) The mean free path is

$$\lambda = \frac{1}{n\sigma\sqrt{2}} = \left(\frac{1}{n}\right)\frac{1}{(\pi D^2)\sqrt{2}} = \left(\frac{V}{N}\right)\frac{1}{(\pi D^2)\sqrt{2}} = \left(\frac{kT}{p}\right)\frac{1}{(\pi D^2)\sqrt{2}}$$

Therefore,

$$\lambda = \left[\frac{(1.38\times10^{-23} \frac{J}{K})(300 \text{ K})}{1.3\times10^{-7}\text{Pa}}\right]\frac{1}{\pi(10^{-10} \text{ m})^2 \sqrt{2}} = \boxed{7.0\times10^5 \text{ m}}$$

(b) Because the mean free path is inversely proportional to the pressure, we have

$$\lambda_2/\lambda_1 = p_1/p_2 = (10^{-9} \text{ torr})/(1.0 \times 10^{-6} \text{ torr}) = 1.0 \times 10^{-3}.$$

$\boxed{\text{The mean free path decreases by a factor of 1000 to } 7.0 \times 10^2 \text{ m}}$.

Answers to Practice Quiz

1. (d) 2. (e) 3. (d) 4. (b) 5. (b)

Answers to Practice Problems

1. 836 m/s   2.   1023 m/s
3. 1,959,897 N/m^2 4.   1,780,643 N/m^2
5. 5.2 nm   6.   0.052 eV
7. 600 m/s   8.   12.8 m/s
9. 11.8 m/s   10.   89.9 meV

CHAPTER 20

THE SECOND LAW OF THERMODYNAMICS

Chapter Objectives

After studying this chapter, you should

1. understand the need for the second law of thermodynamics.
2. know several equivalent forms of the second law of thermodynamics.
3. know the basic thermodynamic principles behind the workings of heat engines.
4. understand the lessons of the Carnot cycle.
5. be able to determine the efficiencies of non Carnot engines.
6. be able to calculate the change in entropy during a thermodynamic process.
7. understand the meaning of entropy and its relationship to probability.

Chapter Review

In this final chapter on thermodynamics, we focus on the second law of thermodynamics. This law introduces the concept of entropy which plays a very important role in our understanding of the behavior of many thermodynamics systems.

20–1—20–2 Beyond Energy Conservation and The Second Law of Thermodynamics

As discussed previously, the first law of thermodynamics is really a restatement of the conservation of energy. However, it turns out that thermal systems behave in ways that the conservation of energy alone cannot explain. The two principal examples are (a) when two systems are brought into thermal contact, the net heat flow is always from the system at higher temperature to the system at lower temperature; and (b) it is impossible to convert all of the thermal energy in a system to mechanical work. This last statement is equivalent to saying that the efficiency of an engine, η, is always less than 1, where

$$\eta \equiv \frac{W}{Q_h}$$

in which W is the mechanical work done by the engine and Q_h is the heat supplied. These facts show us the need for the second law of thermodynamics.

The second law of thermodynamics can be stated in many ways, here we consider two equivalent forms of this law:

> The Kelvin Form
>
> It is impossible to construct an engine that converts thermal energy into mechanical work with 100% efficiency.

> The Clausius Form
>
> It is impossible to construct an engine whose sole effect is to transfer thermal energy from a lower temperature body to a body at higher temperature.

20–3—20–4 The Carnot Cycle and Other Types of Engines

The second law of thermodynamics tells us that the efficiency of all engines is limited, but how to determine that limit is not stated in either the Kelvin or Clausius forms. The maximum possible efficiency that operates between two temperatures, T_c and T_h, is attained in the following 4-step cycle:

Isothermal expansion → Adiabatic expansion → Isothermal compression → Adiabatic compression

This cycle is known as the Carnot cycle. Each step is performed in such a way that thermal equilibrium is always maintained, that is, each step is reversible. The efficiency of a Carnot engine, called the Carnot efficiency η_C, is given by

$$\eta_C = 1 - \frac{Q_c}{Q_h} = 1 - \frac{T_c}{T_h}$$

where Q_h is the heat flow into the system from the hot reservoir during the isothermal expansion and Q_c is the heat flow from the thermal system into the cold reservoir during the isothermal compression. Any real engine that operates between the same two temperatures, in which the processes cannot be exactly reversible, will have an efficiency less than η_C.

The engines we have been discussing thus far are called heat engines because they use the natural tendency for heat to flow from hot to cold as a means of generating work. A **refrigerator** is basically the reverse of a heat engine. In a refrigerator, work is the input that ultimately leads to the cold reservoir (the refrigerator) becoming even cooler, and a hot reservoir (the room) becoming even warmer. The definition of efficiency given in the previous section is not as useful for a refrigerator. Instead, the effectiveness of a refrigerator is indicated by its **coefficient of performance**, K_{ref}, which compares the heat absorbed from the cold reservoir, Q_{abs}, to the amount of work required to absorb it, W:

$$K_{ref} \equiv \frac{Q_{abs}}{W}$$

A **heat pump** is the same as a refrigerator, except it has a different purpose. The goal of a refrigerator is to make the cold reservoir colder, while the goal of a heat pump is to make the hot reservoir hotter. The coefficient of performance for a heat pump, COP, compares the heat rejected from the pump into the hot reservoir, Q_{rej}, to the amount of work required during the cycle, W:

$$COP = \frac{Q_{rej}}{W}$$

Example 20–1 A heat engine generates 4.1 kJ of work at 35% efficiency. How much heat is exhausted in the cold reservoir?

Setting it up:

The diagram shows a schematic of the heat engine.

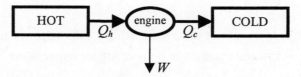

The information provided in the problem is the following:

Given: $W = 4.1$ kJ, $\eta = 0.35$; **Find**: Q_c

Strategy

Because both the efficiency and the work done can be written in terms of Q_h, we can combine those two expressions to eliminate Q_h and solve for Q_c.

Working it out:

By the conservation of energy, the work done is

$$W = Q_h - Q_c \quad \Rightarrow \quad Q_c = Q_h - W$$

We can make use of the efficiency by noting that

$$\eta = \frac{W}{Q_h} \quad \Rightarrow \quad Q_h = \frac{W}{\eta}$$

Putting these facts together gives

$$Q_c = \frac{W}{\eta} - W = W\left(\frac{1-\eta}{\eta}\right)$$

Obtaining the numerical result gives

$$Q_c = \left(4.1 \times 10^3 \text{ J}\right)\left(\frac{1-0.35}{0.35}\right) = 7.6 \text{ kJ}$$

What do you think? Based on the above results, would you consider this to be a good heat engine?

Practice Quiz

1. If a heat engine requires 4500 J of heat to produce 1200 J of work, what is its efficiency?
    (a) 0.27   **(b)** 100%   **(c)** 0.73   **(d)** 0.36   **(e)** 0%

2. In terms of process and function, which pair of devices do you think to be most alike?
    (a) heat engine and refrigerator
    (b) heat engine and heat pump
    (c) refrigerator and air conditioner
    (d) air conditioner and heat pump
    (e) air conditioner and heat engine

3. If heat flow to the hot reservoir increases over time for a refrigerator (assuming Q_{abs} remains constant), its coefficient of performance
    (a) increases
    (b) stays the same
    (c) decreases

20–5—20–7 Entropy and the Second Law of Thermodynamics

The fact that heat naturally flows from systems at higher temperature to systems at lower temperature is only one part of a larger "directionality" to the laws of thermodynamics. The basic quantity in physics that encompasses that directionality is called **entropy**, S. The entropy of a system is a state function, and can be defined by how much it changes during a reversible process from one state, A, to another, B, at a fixed temperature T

$$S(B) - S(A) = \int_A^B \frac{dQ}{T}$$

where dQ is a differential heat flow into the system. This definition implies that over any complete reversible cycle, such as a Carnot cycle, $\Delta S = 0$.

In an irreversible process there are losses that reduce the heat flow to the system, $dQ_{irrev} < dQ_{rev}$. This would reduce the value of the above integral if the transformation from state A to state B took place irreversibly. This implies that in general

$$\int_{A \atop irrev}^B \frac{dQ}{T} < \int_{A \atop rev}^B \frac{dQ}{T}$$

Because the right-hand side of the above inequality is the difference in entropy between states A and B, we can also write that inequality as

$$S(B) - S(A) > \int_{A \atop irrev}^B \frac{dQ}{T}$$

The concept of entropy becomes particularly important when we consider the total change in entropy of a system *and its surroundings* during a process. It is found that although the entropy of an individual system can decrease, this total entropy never decreases. For reversible processes, the change in the total entropy is zero, and for all other processes the total entropy increases. Remembering that reversibility is an idealization and that all real processes are irreversible, we see that in reality *the total entropy of the universe is always increasing*. This last point is the sense in which entropy provides us with directionality to the laws of physics — an arrow of time. The statement that the entropy of the universe is increasing is yet another equivalent way to state the second law of thermodynamics. In fact, the second law of thermodynamics is very commonly called "the law of increase of entropy."

Because entropy is a state function, the change in entropy between any two states will be the same regardless of the specific process by which the state of the system change. So, in order to calculate the change in entropy, even during an irreversible process, we only need to imagine a reversible process that will take the system between those same two states, and calculate the change in entropy for that reversible process. As a concrete example, the change in entropy between two states of an ideal gas is given by

$$S(B) - S(A) = C_V \ln \frac{T_B}{T_A} + nR \ln \frac{V_B}{V_A} = C_V \ln \frac{p_B}{p_A} + C_p \ln \frac{V_B}{V_A}$$

One can gain some intuitive insight into the physical meaning of entropy by thinking of it as a measure of the amount of disorder in the universe. When processes occur, the universe always comes away more disordered than it was before. An example is the natural flow of heat from a hot reservoir to a cold reservoir. In the cold reservoir, the molecules are moving more slowly and scattering off each other at a slower rate. If the system is cold enough they may even clump together and solidify, forming a crystal lattice. When heat from the hot reservoir comes in, it increases the speed of the molecules, causing them to move about more randomly and increasing the disorder (and therefore the entropy) of the system. Of course, the disorder (entropy) of the hot reservoir decreases, but it can be shown that the increase in entropy of the cold reservoir is always greater in magnitude than the decrease in entropy of the hot reservoir. The total disorder always increases. All of the consequences of the second law of thermodynamics can be accounted for by treating the entropy as a measure of the amount of disorder in a system.

Example 20–2 A mass of 2.88 kg of ice at 0.00 °C is mixed with 0.500 kg of steam at 100.0 °C. Estimate the change in entropy only up to the point when all the ice has melted and all the steam has condensed.

Setting it up:

The information provided in the problem is the following:

Given: m_{ice} = 2.88 kg, $t_{0,ice}$ = 0.00 °C, m_{steam} = 0.500 kg, $t_{0,steam}$ = 100.0 °C;   **Find:** ΔS

Strategy:

To estimate the total change in entropy, we need to calculate the heat flow for all the processes that take place until we reach the point of having all liquid water.

Working it out:

From chapter 18, we know the latent heat of fusion for water, L_f = 79.6 cal/g. So, we can determine the heat flow into the ice needed to melt it from

$$\Delta Q_{ice} = m_{ice} L_f = \left(2.88 \times 10^3 \text{g}\right)\left(79.6 \text{ cal/g}\right) = 2.29248 \times 10^5 \text{ cal}$$

The resulting change in entropy is

$$\Delta S_{ice} = \frac{\Delta Q_{ice}}{T_{0,ice}} = \frac{2.29248 \times 10^5 \text{cal}\left(\dfrac{4.185 \text{ J}}{\text{cal}}\right)}{(0.00 + 273.15) \text{ K}} = 3512 \text{ J/K}$$

Similarly, we know the latent heat of vaporization for water, L_v = 540 cal/g. Therefore, the heat flow *into* the steam needed to condense it is given by

$$\Delta Q_{steam} = -m_{steam} L_v = -\left(0.500 \times 10^3 \text{g}\right)\left(540 \text{ cal/g}\right) = -2.70000 \times 10^5 \text{cal}$$

The resulting change in entropy is

$$\Delta S_{steam} = \frac{\Delta Q_{steam}}{T_{0,steam}} = \frac{-2.70000 \times 10^5 \text{cal}\left(\dfrac{4.185 \text{ J}}{\text{cal}}\right)}{(100.0 + 273.15) \text{ K}} = -3028 \text{ J/K}$$

Given that $|\Delta Q_{steam}| > \Delta Q_{ice}$, we know that there is heat that flows from the condensing steam into the melted ice in an amount $\Delta Q_{liquid} = |\Delta Q_{steam}| - \Delta Q_{ice} = 4.0752 \times 10^4 \text{cal}$. The change in temperature of the liquid that comes from the melted ice is given by

$$\Delta Q_{\text{liquid}} = m_{\text{ice}} c_{\text{water}} \Delta T_{\text{liquid}} \quad \Rightarrow \quad \Delta T_{\text{liquid}} = \frac{\Delta Q_{\text{liquid}}}{m_{\text{ice}} c_{\text{water}}} = \frac{4.0752 \times 10^4 \, \text{cal}}{(2880 \, \text{g} \times 1.00 \, \text{cal/g} \cdot \text{K})} = 14.15 \, \text{K}$$

To calculate the change in entropy of the liquid, we have

$$\Delta S_{\text{liquid}} = \int_{T_0}^{T_f} \frac{dQ_{\text{liquid}}}{T} = \int_{T_0}^{T_f} \frac{m_{\text{ice}} c_{\text{water}} dT}{T} = m_{\text{ice}} c_{\text{water}} \int_{T_0}^{T_f} \frac{dT}{T} = m_{\text{ice}} c_{\text{water}} \ln\left(\frac{T_f}{T_0}\right)$$

In kelvins, the initial temperature is 273.15 K and the final temperature is 273.15 + 14.15 K = 287.30 K. This tells us that

$$\Delta S_{\text{liquid}} = (2.88 \, \text{kg})[4185 \, \text{J/(kg} \cdot \text{K)}] \ln\left(\frac{287.30 \, \text{K}}{273.15 \, \text{K}}\right) = 608.7 \, \text{J/K}$$

Therefore, we can finally conclude that the total change in entropy is

$$\Delta S_{\text{tot}} = \Delta S_{\text{ice}} + \Delta S_{\text{steam}} + \Delta S_{\text{liquid}} = 3512 \, \text{J/K} - 3028 \, \text{J/K} + 608.7 \, \text{J/K} = 1090 \, \text{J/K}$$

What do you think? Can you determine the total change in entropy for the system to reach its final equilibrium state?

Practice Quiz

4. Can the entropy of a system ever decrease?
    (a) no, because entropy always increases
    (b) yes, but only in some reversible processes
    (c) no, because the change in entropy is zero for any system
    (d) yes, because entropy always decreases
    (e) yes, if there's a net heat flow out of the system

5. When the total entropy increases during a process, the systems involved become
    (a) more orderly
    (b) less energetic
    (c) crystallized
    (d) more disordered
    (e) more energetic

Reference Tools and Resources

I. Key Terms and Phrases

the second law of thermodynamics the thermodynamic law that governs the behavior of entropy

engine a set of thermal transformations that form a cycle

efficiency the fraction of the heat flow that is converted into mechanical work

Carnot cycle a four step cycle that makes up the most efficient engine operating between two given temperatures

heat pump removes energy from a cooler region in order to transfer some of it to a warmer region

refrigerator removes energy to cool a cooler region and transfers it to a warmer region

coefficient of performance alternative measure of efficiency for heat pumps and refrigerators

entropy the ratio of the heat that flows at a fixed temperature to that temperature for a reversible process. It measures the amount of disorder in a system

II. Important Equations

Name/Topic	Equation	Explanation
Efficiency	$\eta \equiv \dfrac{W}{Q_h}$	The efficiency of an engine that does an amount of work W from an amount of heat Q_h.
Carnot cycle	$\eta_C = 1 - \dfrac{T_c}{T_h}$	The Carnot efficiency is the maximum possible efficiency between two temperatures.
Heat pumps	$\text{COP} = \dfrac{Q_{\text{rej}}}{W}$	The performance level of a heat pump that takes an amount of work W and produces an amount of heat Q_{rej}.
Refrigerators	$K_{\text{ref}} = \dfrac{Q_{\text{abs}}}{W}$	The performance level of a refrigerator that takes an amount of work W and absorbs an amount of heat Q_{abs}.
Entropy	$dS = \dfrac{dQ}{T}$	The change in entropy when an amount of heat dQ flows at temperature T in a reversible process.
Entropy	$\Delta S = C_V \ln\left(\dfrac{T_B}{T_A}\right) + nR \ln\left(\dfrac{V_B}{V_A}\right)$	The change in entropy for an ideal gas during a reversible transformation.

III. Know Your Units

Quantity	Dimension	SI Unit
Efficiency (η)	dimensionless	—
Coefficient of performance (COP, K)	dimensionless	—
Entropy (S)	$\left[ML^2T^{-2}K^{-1}\right]$	J/K

Practice Problems

1. An Otto engine (see problem 35 at the end of the chapter) with a diatomic gas as the working substance has a compression ratio of 12. To the nearest percent, what is its efficiency?

2. The "exhaust" temperature of a Carnot Engine is O °C. To the nearest °C, what would its high temperature need to be if it is to have the same efficiency as the answer to problem 1?

3. A heat engine absorbs 4395 J of heat at a temperature of 1091 K and exhausts 1804 J of heat at a temperature of 300 K. To the nearest tenth of a J/K, what is the total change in entropy in this process?

4. A refrigerator removes 943 J of heat from a temperature of 280 K and deposits 1473 J of heat into a room at 310 K. To the nearest single decimal place, what is its COP (Coefficient of Performance)?

5. What would the answer to Problem 4 be if the device were a heat pump moving heat from the outside to the inside of a house?

6. To the nearest hundredth of a J/K, what is the overall change in entropy of the process in Problem 5?

7. 5.1 moles of an ideal gas undergo an adiabatic free expansion from 93 L to 483 L. To the nearest tenth of a J/K, what is the change in entropy?

8. One kilogram of water is taken from 18 °C to 86 °C. To the nearest J/K, estimate the change in entropy of the water.

9. To the nearest J/K, what is the change in entropy of 629 grams of water at 0 °C when it is frozen into ice at 0 °C?

10. If the water in problem 9 were frozen by being placed in contact with a temperature reservoir at − 39 °C, to the nearest J/K, what would be the total entropy change of the reservoir and water?

Selected Solutions to End of Chapter Problems

11. (a) The maximum efficiency is
$$\eta = 1 - T_c/T_h$$
Solving for T_c gives
$$T_c = T_h(1-\eta)$$
The temperature of the hot reservoir, in kelvin, is $T = (450 + 273.15)\,\text{K} = 723.15\,\text{K}$. Therefore,
$$T_c = T_h(1-\eta) = (723.15\,\text{K})(1-0.55) = \boxed{325\,\text{K}\ \ (52\,°\text{C})}$$

(b) We find the heat flow into the system from the hot reservoir from

$$Q_h = W/\eta = 5.0 \text{ kWh}/0.55 = \boxed{9.1 \text{ kWh}}.$$

We find the heat flow rejected to the cold reservoir from energy conservation:

$$W = Q_h - Q_c \quad \Rightarrow \quad 5.0 \text{ kWh} = 9.1 \text{ kWh} - Q_c$$

which gives $Q_c = \boxed{4.1 \text{ kWh}}$.

29. First we check to see if the first law of thermodynamics is satisfied. The net heat transfer is

$$(dQ/dt)_h - (dQ/dt)_c = 94 \text{ cal/s} - 80 \text{ cal/s} = 14 \text{ cal/s}.$$

The rate at which work is done by the compressor is

$$P = dW/dt = (60 \text{ W})/(4.185 \text{ J/cal}) = 14 \text{ cal/s}$$

so the law is satisfied.

We check to see if the second law of thermodynamics is satisfied. The coefficient of performance of the refrigerator is

$$K_{\text{ref}} = \frac{(dQ/dt)_h}{dW/dt} = \frac{94 \text{ cal/s}}{14 \text{ cal/s}} = 6.7$$

The ideal coefficient of performance for these temperatures is

$$K_{\text{ref}} = T_c/(T_h - T_c) = (253 \text{ K})/(298 \text{ K} - 253 \text{ K}) = 5.6$$

Because this is less than the specifications and actual coefficients of performance will be significantly less than ideal, the statement $\boxed{\text{cannot be trusted}}$.

41. For the system of iron and water, the net heat flow is zero:

$$Q_{\text{iron}} + Q_{\text{water}} = m_{\text{iron}}c_{\text{iron}} \Delta T_{\text{iron}} + m_{\text{water}}c_{\text{water}} \Delta T_{\text{water}} = 0$$

This can be rewritten as

$$m_{\text{iron}}c_{\text{iron}} \left(T - T_{0,\text{iron}} \right) + m_{\text{water}}c_{\text{water}} \left(T - T_{0,\text{water}} \right) = 0$$

Collecting terms gives,

$$\left(m_{\text{iron}}c_{\text{iron}} + m_{\text{water}}c_{\text{water}} \right) T - m_{\text{iron}}c_{\text{iron}}T_{0,\text{iron}} - m_{\text{water}}c_{\text{water}}T_{0,\text{water}} = 0$$

Solving this for the final temperature T gives

$$T = \frac{m_{\text{iron}}c_{\text{iron}}T_{0,\text{iron}} + m_{\text{water}}c_{\text{water}}T_{0,\text{water}}}{\left(m_{\text{iron}}c_{\text{iron}} + m_{\text{water}}c_{\text{water}} \right)}$$

Before we can determine the final temperature, we need the mass of water. This is found by

$$0.5 \text{ L} = 0.5 \times 10^3 \text{ cm}^3 \quad \Rightarrow \quad m_{\text{water}} = \left(0.5 \times 10^3 \text{ cm}^3 \right)\left(1.0 \text{ g/cm}^3 \right) = 0.5 \times 10^3 \text{ g}$$

Thus, the equilibrium temperature is

$$T = \frac{\left(10^3 \text{ g} \right)\left(0.107 \tfrac{\text{cal}}{\text{g·K}} \right)\left(80°\text{C} \right) + \left(0.5 \times 10^3 \text{ g} \right)\left(\tfrac{1 \text{ cal}}{\text{g·K}} \right)\left(20°\text{C} \right)}{\left(10^3 \text{ g} \right)\left(0.107 \tfrac{\text{cal}}{\text{g·K}} \right) + \left(0.5 \times 10^3 \text{ g} \right)\left(\tfrac{1 \text{ cal}}{\text{g·K}} \right)} = \boxed{31°\text{C}}$$

We calculate the entropy change for each component from a reversible path, where it is in contact with an infinite series of reservoirs:

$$\Delta S = \int \frac{dQ}{T} = \int mc \frac{dT}{T} = mc \ln\left(\frac{T_f}{T_i}\right)$$

Therefore,

$$\Delta S_{iron} = (10^3 \text{ g})(0.107 \text{ cal/g·K})(4.185 \text{ J/cal}) \ln (304 \text{ K}/353 \text{ K}) = -67 \text{ J/K}$$

$$\Delta S_{water} = (0.50 \times 10^3 \text{ cm}^3)(1.0 \text{ g/cm}^3)(1 \text{ cal/g·K})(4.185 \text{ J/cal}) \ln (304 \text{ K}/293 \text{ K}) = +77 \text{ J/K}$$

The total entropy change is

$$\Delta S = \Delta S_{iron} + \Delta S_{water} = \boxed{+ 10 \text{ J/K}}.$$

51. In an isothermal compression of an ideal gas there is no change in internal energy but there is work done on the gas, which is a negative work done by the gas. From the first law of thermodynamics, there must be a heat flow from the gas: $Q < 0$. The entropy change is

$$\Delta S = Q/T < 0.$$

The entropy of the gas $\boxed{\text{decreases}}$. Note that the entropy of the surroundings increases.

59. Because the internal energy depends only on the temperature, for the isothermal process $\Delta U = 0$, and for a compression $W < 0$. Therefore, the first law of thermodynamics tells us that $Q = W < 0$. The entropy change of the gas is, therefore,

$$\Delta S_{gas} = \boxed{Q/T < 0 \text{ (decreases)}}.$$

The entropy change of the reservoir is

$$\Delta S_{reservoir} = \boxed{-Q/T > 0 \text{ (increases)}}.$$

For the universe, we have $\Delta S_{universe} = \Delta S_{gas} + \Delta S_{reservoir} = 0$.

There is $\boxed{\text{no conflict}}$ with the second law of thermodynamics, because this is a $\boxed{\text{reversible process}}$.

75. The maximum efficiency is

$$\eta_{max} = 1 - T_c/T_h = 1 - 400 \text{ K}/750 \text{ K} = 0.467.$$

The operating efficiency is

$$\eta = 0.60 \eta_{max} = W/Q_h = P_{output}/(dQ_h/dt)$$

Solving for the rate of heat flow from the hot reservoir gives

$$\frac{dQ_h}{dt} = \frac{P_{output}}{0.60\eta_{max}} = \frac{300 \text{ MW}}{0.60(0.467)} = 1070 \text{ MW}$$

We find the waste heat from

$$P_{output} = (dQ_h/dt) - (dQ_c/dt)$$

$$300 \text{ MW} = 1070 \text{ MW} - (dQ_c/dt)$$

which gives $dQ_c/dt = \boxed{770 \text{ MW}}$.

We find the required water flow from

$$dQ_c/dt = (dm/dt)c\,\Delta T \quad \Rightarrow \quad \frac{dm}{dt} = \frac{dQ_c/dt}{c\,\Delta T}$$

Therefore,

$$\frac{dm}{dt} = \frac{770 \times 10^6\,\text{W}}{\left(1\frac{\text{cal}}{\text{g·K}}\right)\left(\frac{4.185\,\text{J}}{\text{cal}}\right)\left(\frac{1000\,\text{g}}{\text{kg}}\right)\left(12\,^\circ\text{C}\right)} = \boxed{1.5 \times 10^4\,\text{kg/s}}$$

Answers to Practice Quiz

1. (a) **2.** (c) **3.** (c) **4.** (e) **5.** (d)

Answers to Practice Problems

1. 63%   **2.** 465 °C

3. −2.0 J/K   **4.** 1.8

5. 2.8   **6.** 1.38 J/K

7. 69.9 J/K   **8.** 879 J/K

9. −767 J/K   **10.** 128 J/K